BASIC NUCLEAR ELECTRONICS

BASIC NUCLEAR

ELECTRONICS

Hai Hung Chiang

Electronic Engineer, IAEA Fellow
Argonne National Laboratory

WILEY-INTERSCIENCE / A DIVISION OF JOHN WILEY & SONS

New York • London • Sydney • Toronto

Library of Congress Catalog Card Number 77-77836
SBN 471 15503 9 Printed in the United States of America

FOREWORD

Modern nuclear physics experiments rely heavily on radiation detectors and their associated electronic circuits for data acquisition. In addition, there is a rapidly growing trend toward the use of electronic data processing and presentation. Large nuclear-electronic systems such as multichannel analyzers and on-line computers require many components; hence high reliability is a necessity. The transistor has made reliability, small size, and economy possible. The transition from tubes to transistors is essentially complete, whereas the transition from discrete transistors and components to integrated circuits is currently being made.

Mr. Hai Hung Chiang saw the need for bridging the "gaps of understanding" between the nuclear radiation detector and its data output, between the older vacuum-tube theory of nuclear-electronic circuits and the newer transistorized or integrated circuit forms, and, finally, between textbook circuits and the circuits found in laboratory practice. This book does a good job of bridging these gaps.

Electronics Division
Argonne National Laboratory
August 1968

William W. Managan
Associate Director

PREFACE

Recognizing the need for a fundamental presentation of basic nuclear electronics, Mr. Hai Hung Chiang set out early in 1967 to bridge the sizable gap between mathematics and application. The book contains four chapters. Chapter 1 is subdivided into twenty-one sections, Chapter 2, into seven sections, Chapter 3 into four sections, and Chapter 4, into six sections. The book will be of value in the college classroom, in the laboratory, and as a reference manual for engineers in industry.

The first chapter is a basic introduction to the resistor, capacitor, and inductor and discusses their interconnection for high-pass and low-pass filters. In addition, it provides a good introduction to tubes and transistors when used in basic logic circuits and amplifiers, both dc and ac.

With Chapter 1 as a foundation, Mr. Chiang brings to the reader in Chapter 2 the basics of pulse amplifiers and single-channel analyzers. This includes information on the practical reasons for various sources of pulse-height distortion. With good diagrams and formulas, typical circuits of low-noise preamplifiers for semiconductor radiation detectors are presented.

The material in Chapter 3 increases the scope of the book to a systems-engineering category. Complete systems, such as multichannel pulse-height analyzers and analog-to-digital converters, are discussed and treated deeply enough to preclude supplementary reading.

Chapter 4 discusses the various gamma-ray and charged-particle detectors and their associated counting circuits. An integrated-circuit six-decade scaler is presented in exacting detail.

Since the organization of the material and style of presentation are original with the author, acknowledgments of contributions of others are best left to a comprehensive bow in the direction of each entry in the bibliography and to the Electronics Division engineers at Argonne National Laboratory.

Dr. James P. Bobis
Associate Electrical Engineer

Electronics Division
Argonne National Laboratory
August, 1968

AUTHOR'S PREFACE

The purpose and content of *Basic Nuclear Electronics* have been outlined by Dr. J. P. Bobis in the preceding preface. The techniques of nuclear-electronic instrumentation are rapidly expanding, and it was necessary that a book that bridged the gap between theory and practice be written.

The basic circuits related to nuclear instrumentation are systematically described in Chapter 1. If the reader is familiar with them, he can study and understand the succeeding typical circuitries by himself by considering them as design examples. It is recommended that Chapter 1 be carefully studied before the student proceeds to succeeding chapters of the book.

It is a pleasure to recognize and express appreciation for the assistance given by many persons of the Argonne National Laboratory. The foreword and preface for this book were done by Mr. William W. Managan, Associate Director, and Dr. James P. Bobis, Associate Electrical Engineer, Electronics Division, and some equations were derived by Dr. Bobis. Reference materials and some explanation of them were furnished by Messrs. W. W. Managan, J. P. Bobis, M. G. Strauss, I. S. Sherman, O. D. Despe, J. M. Paul, D. J. Keefe, W. P. McDowell, S. J. Rudnick, E. L. Williams, L. L. Sifter, E. W. Johanson, R. Brenner, G. D. Ansley,

T. W. Yannitell, J. E. Miranda, J. G. Ello, B. Zabransky, Dr. Stanley Ruby, and Dr. M. J. Oestmann. Assistance was provided by Messrs. G. F. Meravi, J. A. Redmond, W. R. Erdman, G. J. Kamis, and R. R. Whitman, and corrections of the first draft were made by Messrs. D. J. Keefe, W. P. McDowell, J. M. Paul, E. L. Williams, O. D. Despe, W. K. Brookshier, A. L. Winiecki, and J. L. Blahunka. Without their cooperation and competent assistance this book could not have been completed.

Hai Hung Chiang

Electronics Division
Argonne National Laboratory
August 1968

CONTENTS

Chapter 1 BASIC ELECTRONIC DEVICES AND CIRCUITS   1

1-1 Linear Circuit Elements   1
1-2 Basic Network Theorems   4
1-3 The High-pass *RC* Network   5
1-4 The Low-pass *RC* Network   16
1-5 *RLC* Circuits   23
1-6 Pulse Transformers   28
1-7 Delay Lines   38
1-8 Introduction to Diodes   43
1-9 Basic Diode Circuits   51
1-10 Characteristics of Vacuum Triode Tubes   62
1-11 Transistors   64
1-12 Low-frequency Small-Signal Transistor Model   71
1-13 The Hybrid-π Common-Emitter Model – the
     High-frequency Small-signal Equivalent Circuit   77
1-14 Operational Amplifiers – One Important Application
     of DC Amplifiers   80
1-15 Some Other Applications of DC Amplifiers   85
1-16 Switching Circuits   88
1-17 Basic Logic Circuits 100

1-18 Transistor Oscillators 117
1-19 Flip-flop Circuits 125
1-20 Monostable Multivibrators 144
1-21 Binary Adders 146

Chapter 2   PULSE AMPLIFIERS AND SINGLE-CHANNEL
ANALYZERS   155

2-1 Operation of Current-mode and Voltage-mode Signals 155
2-2 Various Sources of Pulse-height Distortion 160
2-3 Basic Preamplifiers 173
2-4 Feedback Loops 181
2-5 Low-Noise Preamplifiers for Semiconductor
    Radiation Detectors 183
2-6 Typical Circuits for the Main Linear Amplifiers 189
2-7 Single-Channel Pulse-height Analyzers 190

Chapter 3   MULTICHANNEL PULSE-HEIGHT ANALYZERS 215

3-1 Multichannel Pulse-height Analysis 215
3-2 Introduction to the Typical One-Parameter Analyzers 225
3-3 Analog-to-Digital Converter (ADC) Circuits 235
3-4 Ferrite Core Memories 252

Chapter 4   TYPICAL NUCLEAR AND ELECTRONIC
INSTRUMENTS   259

4-1 Typical G-M Tube Survey Meters 259
4-2 Ion-Chamber Type Survey Meter Model 440 263
4-3 Gas Proportional Alpha Counter Model PAC-4G 269
4-4 Exponential Pulse Generator Model PG-75 277
4-5 Type 321 Oscilloscope 277
4-6 Typical Scaler-Integrated Circuit 303

Index   339

BASIC NUCLEAR ELECTRONICS

CHAPTER 1

BASIC ELECTRONIC DEVICES AND CIRCUITS

1-1 LINEAR CIRCUIT ELEMENTS

Resistors

Resistance is defined as the physical property of an element, device, branch, network, or system that is the factor by which the mean-square conduction current must be multiplied to give the corresponding power lost by dissipation as heat or as other permanent radiation or loss of electromagnetic energy from the circuit. Through any surface, the integral of the normal component of the conduction current density over that surface is referred to as the conduction current. Note that the conduction current is a scalar and hence has no direction. It does not flow, and so is different from the direct current (dc) or alternating current (ac). Direct current is a unidirectional current in which the changes in value are either zero or so small that they may be neglected. Alternating current has positive and negative values during a finite time interval. The resistance (ohm) of a conductor is equal to $\rho(l/A)$, where l is the length in centimeters, A is the cross-section area in square centimeters, and ρ is the resistivity of the material. The resistor is a physical device used

1

primarily to introduce resistance into an electric circuit. Because resistance R opposes the current (dc or ac) flow, a resistor will develop a voltage drop V and dissipate the electric power P as heat if the current I passes through it. Among these quantities the relations are $V = IR$ and $P = IV = I^2R = V^2/R$, where V, I, R, and P are expressed in volts, amperes, ohms, and watts, respectively. If n resistors of $r_1, r_2, r_3, \ldots, r_n$ are connected in series, the total resistance will be $R = r_1 + r_2 + r_3 + \cdots + r_n$. On the other hand, if they are connected in parallel, the total resistance will be $R = 1/(1/r_1 + 1/r_2 + 1/r_3 + \cdots + 1/r_n)$. The sum of the voltage drops across n resistors in series equals the source voltage. The sum of the currents through n resistors in parallel equals the total current. The metal resistance increases with temperature, but the nonmetal resistance decreases with an increase of temperature. The rating of resistors is usually expressed in watts and ohms (or $k\Omega$, $M\Omega$).

Inductors

An inductor (inductance coil, retardation coil, or retard coil) is a device whose primary purpose is to introduce inductance into an electric circuit. Inductance is a number of magnetic flux linkages per ampere, or $L = N\phi/I$, where L is inductance expressed in henrys, I is the current in amperes, ϕ is the magnetic flux in webers (1 weber $= 10^8$ maxwells), N is the number of turns, and $N\phi$ is the number of flux linkages. An inductor stores electric energy in the magnetic field around it. The stored energy is $W_L = LI^2/2$, where W_L, L, and I are expressed in joules, henrys, and amperes, respectively. Because inductance opposes the change of the current passing through the circuit, an inductor will generate a back emf and introduce a reactance in the ac circuit [f is the frequency in cycles per second (Hz) greater than zero]. The back emf is $e = -L$ (change in current) / (change in time) $= -L(di/dt)$ volts. The inductive reactance is $X_L = \omega L = 2\pi fL$ ohms. A mutual inductor is an inductor for changing the mutual inductance between two circuits. If there are inductances L_1 and L_2 in these two circuits, the mutual inductance will be $M = K\sqrt{L_1L_2}$ henrys, where K is the coefficient of coupling between the two circuits and is usually less than unity. If the two inductances L_1 and L_2 are connected in series, the total inductance will be $L = L_1 + L_2 + 2M$ for series aiding and $L = L_1 + L_2 - 2M$ for series opposing. On the other hand, if they are connected in parallel, the total inductance will be $L = 1/[1/(L_1 + M) + 1/(L_2 + M)]$ for parallel aiding, or $L = 1/[1/(L_1 - M) + 1/(L_2 - M)]$ for parallel opposing.

Capacitors

A capacitor is a device whose primary purpose is to introduce capacitance into an electric circuit. Capacitance (capacity, C) is the property of a system of conductors and dielectrics that permits the storage of electrically separated charges when potential differences exist between the conductors. Its value is expressed as the ratio of a quantity of electricity to a potential difference, or $C = q/V$, where C, $q(= It)$ and V are expressed in farads, coulombs, and volts, respectively. The capacitor used in electronic circuits consists of conductors, usually in the form of two plates or other pairs of similar shapes with a large surface area, separated by a dielectric. When a simple capacitor is charging, electrons are accumulating on one plate while the other plate is supplying electrons to the circuit. This capacitor stores the electric energy in the electric field about the capacitor plates. The stored energy is $W_c = CV^2/2$, where W_c, C, and V are expressed in joules, farads, and volts, respectively. Because capacitance opposes the change of the applied voltage, a capacitor will introduce a reactance in the ac circuit. The capacitive reactance is $X_c = 1/(2\pi fC)$, where X_c, f, and C are expressed in ohms, cycles per second (Hz), and farads, respectively. The total capacitance of n capacitors connected in series is $C = 1/(1/c_1 + 1/c_2 + 1/c_3 + \cdots + 1/c_n)$. The total capacitance of n capacitors connected in parallel is $C = c_1 + c_2 + c_3 + \cdots + c_n$. The rating of capacitors is usually expressed in working volts dc and microfarads or picofarads (micro-microfarads). The polarity of polarized capacitors must be observed; otherwise damage or failure will occur.

RLC Used in AC Circuits

In a pure resistive circuit, the current and voltage are in phase. In a pure inductive circuit, the current lags the voltage by 90°. In a pure capacitive circuit, the current leads the voltage by 90°. In the series RLC circuit of Figure 1-1 the impedance (ohms) is $Z = [R^2 + (X_L - X_c)^2]^{1/2}$, the applied voltage is $V = [V_R^2 + (V_L - V_c)^2]^{1/2}$, the current is $I = V/Z$, the power factor is $\cos\theta = R/Z$, and the power dissipated is $P = VI\cos\theta$. If $X_L = X_c$, the circuit will resonate. Under this condition, $Z = R = \text{min}$, and hence $I = V/R = \text{max}$. The resonant frequency is $f_r = 1/(2\pi\sqrt{LC})$, and the voltage relation is $V_L = V_c = QV$, where $Q = X_L/R$.

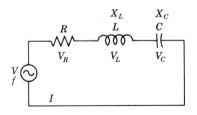

Fig. 1-1 Series *RLC* Circuit

In the parallel *RLC* circuit of Figure 1-2, the impedance is $Z = 1/[(1/R)^2 + (1/X_L - 1/X_c)^2]^{1/2}$, the total current is $I = [I_R^2 + (I_L - I_c)^2]^{1/2}$, the applied voltage is $V = IZ$, the power factor is $\cos \theta = Z/R$, and the power dissipated is $P = VI \cos \theta$.

In the practical parallel resonant circuit of Figure 1-3 R is much less than X_L or X_c. At resonance $X_L = X_c$, $Z = QX_L = $ max, hence $I = (1/Q) \times I_{cir} = $ min, where $Q = X_L/R$.

1-2   BASIC NETWORK THEOREMS

Kirchhoff's Laws

KIRCHHOFF'S CURRENT LAW. The algebraic sum of the currents into and out of each branch point in a given closed network is equal to zero.

KIRCHHOFF'S VOLTAGE LAW. The voltages produced as a result of current(s) flowing in a closed path of a network are equal to the sum of all internal, applied, and induced voltages in the same path.

Superposition Theorem

The response of a linear network containing several independent sources is found by considering each generator separately and then adding the individual responses. When evaluating the response due to one source, one replaces each of the other independent generators by its internal impedance.

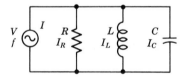

Fig. 1-2   Parallel *RLC* Circuit

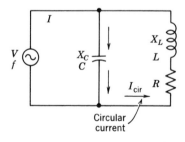

Fig. 1-3.   Practical parallel resonant circuit.

Thevenin's Theorem

Any linear network may, with respect to a pair of terminals, be replaced by a voltage generator in series with the impedance seen at this port (see Figure 1-4*a*). The voltage generator equals the open-circuit voltage between the two terminals.

Norton's Theorem

Any linear network may, with respect to a pair of terminals, be replaced by a current generator in parallel with the impedance seen at this port (see Figure 1-4*b*). The current generator equals the short-circuit current.

1-3 THE HIGH-PASS *RC* NETWORK

Sine-wave Input

A high-pass network is a high-pass filter that attenuates low frequencies and causes a flat-topped pulse to decay. Its simplest form is the *RC* network shown in Figure 1-5*a*. If v_i is a sine-wave generator of frequency f, the output of the network as a function of frequency (see Figure 1-5*b*) is

$$v_0 = \frac{v_i}{[1 + (f_1/f)^2]^{1/2}} \quad \text{and} \quad \theta = \arctan \frac{f_1}{f}, \qquad (1\text{-}1)$$

where $\dfrac{1}{[1 + (f_1/f)^2]^{1/2}} = $ magnitude of gain $|A|$,

$f_1 = $ lower 3-dB frequency $= 1/2\pi RC$,

$\theta = $ angle by which the output leads the input.

At the frequency f_1 the magnitude of the capacitive reactance is equal to the resistance and the gain is 0.707. This drop in signal level corresponds to a signal reduction of 3 dB.[1] The maximum possible gain is equal to 1.

Step-function (Step-voltage) Input

The unit step function $U(t)$ is defined as having the value zero for all negative times and the value unity for all positive times. A step (or step-function) voltage is one which maintains the value zero for all times $t < 0$ and maintains the value V for all times $t > 0$. The response of the network to the step-voltage input is exponential, with a time constant $\tau \equiv RC$, and the output voltage v_0 is of the form

$$v_0 = V_f + (V_i - V_f)\,\epsilon^{-t/\tau}, \qquad (1\text{-}2)$$

[1] $\text{dB} = 10 \log P_2/P_1 = 20 \log V_2/V_1$.

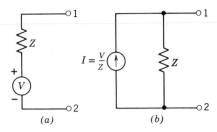

Fig. 1-4 (a) Thevenin equivalent and (b) Norton equivalent for a linear two-terminal network.

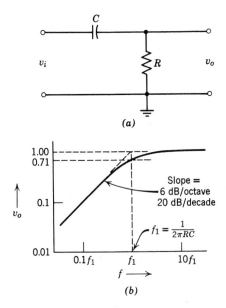

Fig. 1-5 (a) High-pass RC network, (b) Response of the network to a sine-wave input.

where V_f and V_i are the final and initial output voltages, respectively.

There is a principle[1] stating that the voltage across a capacitor cannot change instantaneously provided that the current remains finite. Applying this principle to the network in Figure 1-5a, we must conclude that since at $t = 0$ the input voltage changes discontinuously by an amount V, the output must also change abruptly by this same amount. If we assume that the capacitor is initially uncharged, then the output voltage must jump to

[1] The change in voltage across the capacitor in time t_1 is $(1/C) \int_0^{t_1} i\, dt$. This integral approaches zero as $t_1 \to 0$ provided that the current remains finite.

V immediately after $t = 0$ (at $t = 0+$). Therefore $V_i = V$. For $t > 0$, the input is a constant, and since the capacitor C blocks the dc component of the input, the final output voltage is zero, or $V_f = 0$. Then (1-2) becomes

$$v_0 = V\epsilon^{-t/RC}. \qquad (1-3)$$

The output is 0.61 of its initial value at $0.5RC$, 0.37 at RC, 0.14 at $2RC$, 0.05 at $3RC$, 0.02 at $4RC$, and 0.01 at $5RC$.

Pulse Input

If the pulse in Figure 1-6a is applied to the circuit in Figure 1-5a, the response for times that are less than the pulse duration t_p is the same as that for the step-voltage input. At the end of the pulse, the input falls abruptly by the amount V, and, since the capacitor voltage cannot change instantaneously, the output must also drop by V. Thus immediately after $t = t_p$ (or at $t = t_p+$), $v_0 = V_p - V$; v_0 becomes negative and then decays exponentially to zero, as shown in Figure 1-6b. For $t > t_p$, v_0 is given by

$$v_0 = V(\epsilon^{-t_p/RC} - 1) \; \epsilon^{-(t - t_p)/RC}. \qquad (1-4)$$

For all values of the ratio RC/t_p there must always be an undershoot, and the area below the axis will always equal the area above. Because the input and output are separated by the blocking capacitor C, the dc or average level of the output signal is zero for the high-pass RC network. If $RC >> t_p$, there is only a slight tilt to the output pulse and the undershoot is very small (see Figure 1-6c). If $RC << t_p$, the output consists of a positive spike or pip of amplitude V at the beginning of the pulse and a negative spike of the same size at the end of the pulse, as shown in Figure 1-6d.

Square-wave Input

Figure 1-7a shows a square wave which maintains itself at one constant level V' for a time T_1 and at another constant level V'' for a time T_2 and which is repetitive with a period $T = T_1 + T_2$. For this square-wave input or any other periodic-input waveform, the average level of the steady-state output signal from the circuit of Figure 1-5a is always zero and is independent of the dc level of the input. The output must consequently extend in both the positive and negative directions with respect to the zero-voltage axis, and the area of the part of the waveform above the zero axis must equal the area which is below the zero axis. When the input changes discontinuously by an amount V, the output changes discontinuously by an equal amount and in the same direction. During any finite time interval when the input maintains a constant level, the

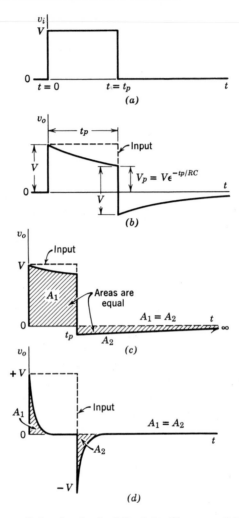

Fig. 1-6   (a) A pulse applied to the circuit of Fig. 1-5a; (b) exponential decay of the output pulse; (c) output pulse if $RC \gg t_p$; (d) output pulse if $RC \ll t_p$.

output decays exponentially toward zero voltage. In the limiting case, where RC/T_1 and RC/T_2 are both arbitrarily large in comparison with unity, the output waveform will be identical to the input, except that the dc component will be lacking (see Figures 1-7a and b). At the other extreme, if RC/T_1 and RC/T_2 are both very small in comparison with unity, the output will consist of alternate positive and negative peaks (see Figure 1-8), and the peak-to-peak amplitude of the output will be twice the peak-to-peak amplitude of the input (see Fig. 1-7a).

More generally, the response to a square wave must appear as in Figure 1-9, where

$$V_1' - V_2 = V, \qquad V_1' = V_1\, \epsilon^{-T_1/RC}, \qquad (1\text{-}5a)$$

$$V_1 - V_2' = V, \qquad V_2' = V_2\, \epsilon^{-T_2/RC}. \qquad (1\text{-}5b)$$

For a symmetrical square wave, $T_1 = T_2 = T/2$, $V_1 = -V_2$ and $V_1' = -V_2'$. Under this condition the equations in (1-5a) are identical with those in (1-5b). Thus we have

$$V_1 = \frac{V}{1 + \epsilon^{-T/2RC}}, \qquad V_1' = \frac{V}{1 + \epsilon^{T/2RC}}. \qquad (1\text{-}6)$$

For $T/2RC \ll 1$ these equations[1] become

$$V_1 \simeq \frac{V}{2}\left(1 + \frac{T}{4RC}\right), \qquad V_1' \simeq \frac{V}{2}\left(1 - \frac{T}{4RC}\right). \qquad (1\text{-}7)$$

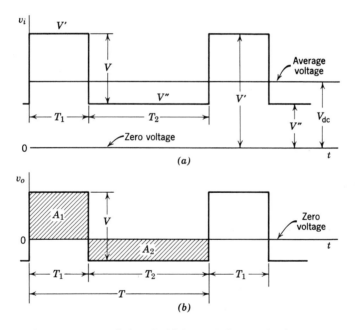

Fig. 1-7 (a) Square wave applied to the high-pass RC network; (b) output waveform if $RC \gg T$. The dc component V_{dc} of the output is always zero. Areas A_1 and A_2 are equal.

[1] Referring to the exponential series, $e^x = 1 + x + x^2/2! + x^3/3! + \cdots$, we have $e^{-T/2RC} = 1 - T/2RC + \cdots$. Hence $V_1 = V/(1 + \epsilon^{-T/2RC}) \simeq V/(2 - T/2RC) \simeq (V/2)\,[1/(1 - T/4RC)] \simeq (V/2)\,[1 \div (1 - T/4RC)] \simeq (V/2)\,(1 + T/4RC)$.

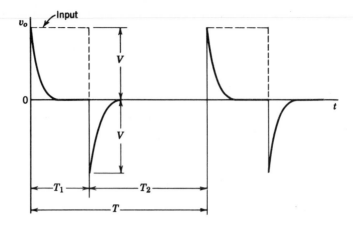

Fig. 1-8   Peaking of a square wave resulting from a very short time constant $(RC \ll T)$.

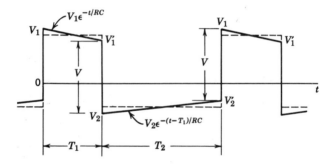

Fig. 1-9   Response of the high-pass RC network to a square-wave input. The dashed curve would represent the output if $RC/T \gg 1$.

Then the exponential portions of the output are approximately linear, as indicated in Figure 1-10. The percentage tilt on the waveform is

$$P_{\text{tilt}} = \frac{V_1 - V_1'}{V/2} \times 100 \simeq \frac{T}{2RC} \times 100 \text{ percent} \qquad (1\text{-}8)$$

or

$$P_{\text{tilt}} = \pi \frac{f_1}{f} \times 100 \text{ percent}, \qquad (1\text{-}9)$$

where $f_1 = 1/(2\pi RC)$, by which the low-frequency 3-dB point is given
$f = $ frequency of the applied square wave $= 1/T$.

Exponential Input

In any RC network, $v_i = q/C + v_0$, where q is the capacitor charge. Differentiating this equation gives

$$\frac{dv_i}{dt} = \frac{i}{C} + \frac{dv_0}{dt} \quad \text{or} \quad \frac{dv_i}{dt} = \frac{v_0}{RC} + \frac{dv_0}{dt}. \tag{1-10}$$

Suppose the input of a high-pass RC network is an exponential waveform given by

$$v_i = V(1 - \epsilon^{-t/\tau}), \tag{1-11}$$

where τ is the time constant of the input wave. Then (1-10) becomes

$$\frac{V}{\tau} \epsilon^{-t/\tau} = \frac{v_0}{RC} + \frac{dv_0}{dt}. \tag{1-12}$$

Defining n and x by

$$n \equiv RC/\tau \quad \text{and} \quad x \equiv t/\tau,$$

the solution[1] of (1-12), subject to the condition that the capacitor voltage is initially zero, is given by

$$v_0 = \frac{nV}{n-1} (\epsilon^{-x/n} - \epsilon^{-x}) \tag{1-13}$$

if $n \neq 1$ and by

$$v_0 = Vx\,\epsilon^{-x} \tag{1-14}$$

if $n = 1$. Equations (1-13) and (1-14) are plotted in Figure 1-11. If $RC \gg \tau$ ($n \gg 1$), then $\epsilon^{-x/n} \gg \epsilon^{-x}$, and (1-13) becomes

$$v_0 \simeq \frac{nV}{n-1} \epsilon^{-x/n} \simeq V\epsilon^{-t/RC}. \tag{1-15}$$

By comparing with (1-3) we see that (1-15) agrees with the way the circuit should behave for an ideal step voltage. Near $t = 0$ the output

---

[1] *Solution by integrating factor.* We first compute the integrating factor $\epsilon^{\int dt/RC} = \epsilon^{t/RC}$. Multiplying (1-12) by this factor yields $(V/\tau)\,\epsilon^{-t/\tau + t/RC} = (v_0/RC)\,\epsilon^{t/RC} + \epsilon^{t/RC}\,(dv_0/dt)$. Notice that the right side of the above equation equals $(d/dt)\,(v_0\,\epsilon^{t/RC})$. Therefore $(V/\tau)\,\epsilon^{[(\tau - RC)/\tau RC]t} = (d/dt)\,(v_0\,\epsilon^{t/RC})$. Integrating we obtain $VRC/(\tau - RC)\int d\,\epsilon^{[(\tau - RC)/\tau RC]t} = \int d(v_0\,\epsilon^{t/CR})$ and $[VRC/(\tau - RC)]\,\epsilon^{[(\tau - RC)/\tau RC]t} = v_0\,\epsilon^{t/RC} + K$ (1-13a) where K is the constant of integration. Substituting the initial condition $v_0 = 0$ at $t = 0$, we obtain $K = VRC/(\tau - RC)$. Now returning to (1-13a), we have $[VRC/(\tau - RC)]\,\epsilon^{[(\tau - RC)/\tau RC]t} = v_0\,\epsilon^{t/RC} + VRC/(\tau - RC)$. Substituting RC by $n\tau$ and t by $x\tau$ and then solving for v_0, we obtain (1-13).

follows the input. Also, the smaller RC is, the smaller the output peak will be (see Figure 1-11).

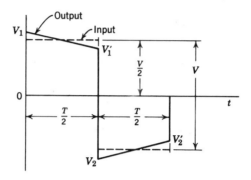

Fig. 1-10   Linear tilt of a symmetry square wave when $RC \gg T$.

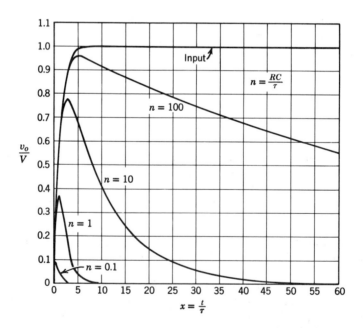

Fig. 1-11   Response of a high-pass RC network to an exponential input. (From J. Millman and H. Taub, *Pulse, Digital, and Switching Waveforms*, McGraw-Hill, New York, 1965.)

Sweep (Ramp) Input

A ramp or sweep voltage is a waveform which is zero for $t < 0$ and which increases linearly with time for $t > 0$, $v = \alpha t$. If the ramp is applied to the network of Figure 1-5a, the output is governed by (1-10), which becomes

$$\alpha = \frac{v_0}{RC} + \frac{dv_0}{dt}. \tag{1-16}$$

Equation 1-16 has the following solution[1] for $v_0 = 0$ at $t = 0$:

$$v_0 = \alpha RC(1 - \epsilon^{-t/RC}). \tag{1-17}$$

For $t << RC$, the exponential in (1-17) may be replaced by a series[2] with the result

$$v_0 = \alpha t(1 - \frac{t}{2RC} + \cdots) \tag{1-18}$$

Figure 1-12a shows the sweep input and output for $RC >> T$. As a measure of the departure from linearity, the transmission error at $t = T$ is defined by

$$e_t \equiv \frac{v_i - v_0}{v_i} \simeq \frac{T}{2RC} = \pi f_1 T, \tag{1-19}$$

where $f_1 = 1/2\pi RC$; for example, if we desire to pass a 1-msec sweep with less than 0.1-percent deviation from linearity, (1-19) yields $RC > 0.5$ sec or $f_1 < 0.3$ Hz.

For $t >> RC$ the output approaches the constant value αRC, as indicated in (1-17) and Figure 1-12b.

RC **Differentiator**

A differentiator is a network in which the output signal is proportional to the derivative of the input signal. One of its simplest forms is a high-pass *RC* network. If, as in Figure 1-5a, the time constant is very small as compared with the time required for the input signal to make an appreciable change, then the network is referred to as a differentiator. In this

---

[1] Solution by integrating factor $\epsilon^{\int dt/RC} = \epsilon^{t/RC}$. Multiplying (1-16) by this factor yields $\alpha\epsilon^{t/RC} = (v_0/RC)\,\epsilon^{t/RC} + \epsilon^{t/RC}\,(dv_0/dt)$. The right side of the above equation equals (d/dt) $(v_0\,\epsilon^{t/RC})$. Therefore $\alpha\epsilon^{t/RC} = (d/dt)\,(v_0\,\epsilon^{t/RC})$. Integrating, we obtain $\int \alpha\epsilon^{t/RC}dt = \int d\,(v_0\,\epsilon^{t/RC})$ and $\alpha RC\,\epsilon^{t/RC} = v_0\,\epsilon^{t/RC} + K$. (1-17a)

Substituting the initial condition $v_0 = 0$ at $t = 0$, we obtain $K = \alpha RC$. Now returning to (1-17a), we have $\alpha RC\epsilon^{t/RC} = v_0\,\epsilon^{t/RC} + \alpha RC$. Solving for v_0, we obtain (1-17).

[2] That is, the exponential series: $e^x = 1 + x + x^2/2! + x^3/3! + x^4/4! + \cdots$.

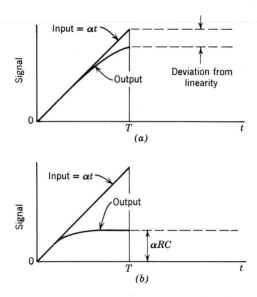

Fig. 1-12 (*a*) response of a high-pass *RC* network to a sweep voltage for $RC \gg T$. (*b*) Response to a sweep voltage for $RC \ll T$; the output approaches αRC after a time has passed corresponding to several times constants.

case, the voltage drop across *R* will be very small relative to the drop across *C*. We may consider that the total input v_i appears across *C*, so that the current is determined entirely by *C*. Therefore the current is $C\, dv_i/dt$, and the output signal across *R* is

$$v_0 = RC\,\frac{dv_i}{dt}. \tag{1-20}$$

Differentiating a square wave gives the waveform which is uniformly zero except at the points of discontinuity. At these points, precise differentiation would yield impulses of infinite amplitude, zero width, and alternating polarity. However, in the actual waveform provided by the *RC* differentiator, the amplitude of the peaks never exceeds *V*, as shown in Figure 1-8. Such an error exists, since, at the time of the discontinuity, the voltage across *R* is not negligible compared with that across *C*.

For the sweep input $v_i = \alpha t$, the output signal is $v_0 = RC\,dv_i/dt = \alpha RC$. This result is true except near $t = 0$ (see Figure 1-12*b*). The error near $t = 0$ is the result of causes similar to those given above.

Consider how to obtain a criterion for good differentiation in terms of steady-state analysis. If a sine wave $v_i = V_m \sin \omega t$ is applied to the circuit of Figure 1-5a, the output will be a sine wave shifted by a leading angle θ such that

$$\tan \theta = \frac{X_c}{R} = \frac{1}{\omega RC} \qquad (1\text{-}21)$$

and the output will be proportional to $\sin(\omega t + \theta)$. In order to obtain $RC\, dv_i/dt = V_m\, \omega RC \cos \omega t$, the angle θ must equal 90°. If $\omega RC = 0.01$, then $1/\omega RC = 100$ and $\theta = 89.4°$, which is sufficiently close to 90° for most purposes. Therefore, if $\omega RC \ll 1$, the output is approximately the expected value $V_m\, \omega RC \cos \omega t$.

The output is often followed by a high-gain amplifier. Any drift in amplifier gain may affect the level of the signal, and amplifier nonlinearity may affect the accuracy of differentiation. These difficulties are avoided by using the operational differentiator (see Sec. 1-14).

The RC differentiator, if used without an integrator in a nuclear pulse amplifier having good high-frequency response, produces a pulse shape like that shown in Figure 1-13. (It is assumed that the input signal is from a detector having a short collection time.) From the standpoint of pulse-height analysis, this pulse shape is very poor.

Double Differentiation

Assume that the amplifier of Figure 1-14 operates linearly and that its output impedance is small compared with the impedance of R_2 and C_2, so that this combination does not load the amplifier. Let R_1 be the parallel combination of R and the input impedance of the amplifier. If the time constants $R_1 C_1$ and $R_2 C_2$ are small compared with the period of the input signal, then this double-RC-clipped amplifier performs approximately a second-order differentiation.

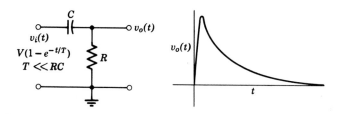

Fig. 1-13   Pulse shape from RC differentiator and fast amplifier.

Consider the exponential waveform $v_i = V(1 - \epsilon^{-t/\tau})$ applied to the circuit of Figure 1-14. If $R_1C_1 = R_2C_2 = \tau$, the output[1] is given by

$$v_0 = -AVx \left(1 - \frac{x}{2}\right) \epsilon^{-x} \tag{1-22}$$

where A is the magnitude of the amplifier gain and $x \equiv t/\tau$. This result is plotted in Figure 1-15. The initial slope of v_0 is $-AV/\tau$, since V/τ is the initial slope of v_i. Note that the output waveform has as much area above the time axis as it does below, which is a fact of importance in pulse spectrometry. However, this area balance will fail if the amplifier is driven out of its linear range.

1-4   THE LOW-PASS RC NETWORK

Sine-wave Input

A low-pass network is a low-pass filter that attenuates high frequencies, causing a rounding of the leading edge of a square pulse. Its simplest form is the RC network shown in Figure 1-16a. If v_i is a sine-wave generator of frequency f, the output of the network as a function of frequency (see Figure 1-16b) is

$$v_0 = \frac{v_i}{[1 + (f/f_2)^2]^{1/2}} \quad \text{and} \quad \theta = -\arctan\frac{f}{f_2}, \tag{1-23}$$

where $\dfrac{1}{[1 + (f/f_2)^2]^{1/2}}$ = magnitude of gain $|A|$

$$f_2 = \text{upper 3-dB frequency} = \frac{1}{2\pi RC}$$

θ = angle θ by which the output leads the input.

The gain falls to 0.707 of its low-frequency value at the frequency f_2.

Step-function (Step-voltage) Input

The response of the low-pass network to a step-function input is exponential with a time constant RC. Because the capacitor voltage

---

[1] The output v of the inverting amplifier (Figure 1-14) is the negative of the waveform in Figure 1-11 and is given by $v = (-AVt/\tau)\epsilon^{-t/\tau}$, where A is the magnitude of the amplifier gain. Now, from (1-10), $dv/dt = v_0/\tau + dv_0/dt$ or $(-AV/\tau)(1 - t/\tau)\epsilon^{-t/\tau} = v_0/\tau + dv_0/dt$. (1-22$a$) Multiplying (1-22$a$) by the integrating factor $\epsilon^{\int dt/\tau} = \epsilon^{t/\tau}$ yields $(-AV/\tau)(1 - t/\tau) = (v_0/\tau)\epsilon^{t/\tau} + (dv_0/dt)\epsilon^{t/\tau}$. The right side of the above equation equals $(d/dt)(v_0\epsilon^{t/\tau})$. Therefore $(-AV/\tau)(1 - t/\tau)dt = d(v_0\epsilon^{t/\tau})$. Integrating, we obtain $(-AVt/\tau)(1 - t/2\tau) = v_0\epsilon^{t/\tau} + K$. (1-22$b$) Substituting the initial condition $v_0 = 0$ at $t = 0$, we obtain $K = 0$. Now returning to (1-22b), we have $(-AVt/\tau)(1 - t/2\tau) = v_0\epsilon^{t/\tau}$. Solving for v_0, we obtain (1-22).

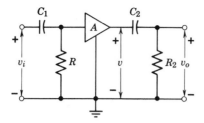

Fig. 1-14 A rate-of-rise amplifier (double-RC-clipped amplifier).

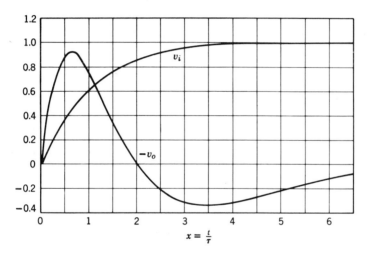

Fig. 1-15 Response of a double differentiator to an exponential input. The numerical values correspond to $A = 4$ and $V = 1$.

cannot change instantaneously, the output starts from zero and rises toward the steady-state value V, as indicated in Figure 1-17. Thus from (1-2) we have

$$v_0 = V(1 - \epsilon^{-t/RC}). \tag{1-24}$$

The pulse rise time t_r is conventionally defined as the time interval between the 10- and 90-percent amplitude points on the wavefront (see Figure 1-17). The time required for v_0 to reach 10 percent of its final value is $0.1RC$ and the time required for it to reach 90 percent of its final value is $2.3RC$. The difference between these two values is the rise time t_r of the network, and it is given by

$$t_r = 2.2RC = \frac{2.2}{2\pi f_2} = \frac{0.35}{f_2}. \tag{1-25}$$

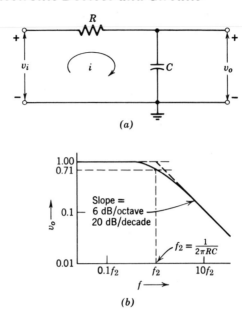

(a)

(b)

Fig. 1-16   (*a*) Low-pass *RC* network; (*b*) response of the network to a sine-wave input.

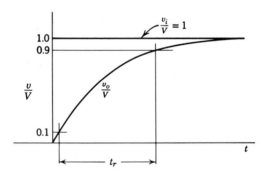

Fig. 1-17   Response of the low-pass *RC* network to a step input.

Pulse Input

The response of the low-pass network to a pulse input for times *t* less than the pulse width t_p is the same as that for a step-function input and is given by (1-24). At $t = t_p$ the voltage is V_p and the output must decrease to zero from this value with a time constant RC, as shown in

Figure 1-18. Note that the output will always extend beyond the pulse width t_p, since whatever charge has accumulated on C during the pulse cannot leak off instantaneously. If it is desired to minimize this distortion, then the rise time must be small relative to the pulse width. A pulse shape will be preserved if the 3-dB frequency f_2 is approximately equal to the reciprocal of the pulse width t_p or if the rise time is $t_r \approx 0.35\ t_p$. Thus, to pass a 0.4-μsec pulse reasonably well requires a circuit with an upper 3-dB frequency of the order of 2.5 MHz or a rise time of 0.14 μsec.

Square-wave Input

A square wave applied to a low-pass RC network is shown in Figure 1-19a. If the time constant RC is small relative to the period T of the input square wave, the output will appear as in Figure 1-19b, which is a reasonable reproduction of the input. If the time constant RC is comparable to the period T, the output will be as in Figure 1-19c. If $RC >> T$, the output consists of exponential sections which are essentially linear (see Figure 1-19d).

By referring to (1-2), one can give the equations for the rising portion and the falling portion in Figure 1-19c. They are

$$v_{o1} = V' + (V_1 - V')\ \epsilon^{-t/RC} \tag{1-26}$$

and

$$v_{o2} = V'' + (V_2 - V'')\ \epsilon^{-(t\,-\,T_1)/RC}. \tag{1-27}$$

These two equations can be solved for the two unknowns[1] V_1 and V_2 if we set $v_{o1} = V_2$ at $t = T_1$ and $v_{o2} = V_1$ at $t = T_1 + T_2$.

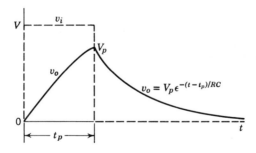

Fig. 1-18 Response of the low-pass RC network to a pulse input.

---

[1] For a symmetrical square wave with zero average value, $T_1 = T_2 = T/2$, $V' = -V'' = V/2$, $V_1 = -V_2$, and $V_2 = (V/2)\ [(\epsilon^{2x} - 1)/(\epsilon^{2x} + 1)] = \dfrac{V}{2}\ \tanh x$, (1-28) where $x \equiv T/4RC$.

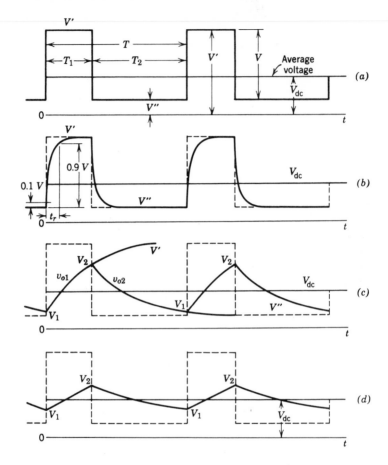

Fig. 1-19 Response of the low-pass *RC* network to a square-wave input. (*a*) Square-wave input; (*b*, *c*, *d*) output voltage. Relative to the period *T*, the time constant is small for (*b*), comparable for (*c*), and large for (*d*). The dc voltage V_{dc} at the output is the same as that of the input.

Exponential Input

For an exponential input of the form of (1-11)

$$v_i = V(1 - \epsilon^{-t/\tau}) \quad \text{or} \quad v_i = V(1 - \epsilon^{-x})$$

the voltage output across the capacitor is the difference between the input given by (1-11) and the voltage v_R across the resistor. For $n \neq 1$, v_R is given by (1-13) and we have

$$\frac{v_o}{V} = 1 + \frac{1}{n-1}\,\epsilon^{-x} - \frac{n}{n-1}\,\epsilon^{-x/n}. \tag{1-29}$$

For $n = 1$, v_R is given by (1-14), and

$$\frac{v_o}{V} = 1 - (1+x)\,\epsilon^{-x}. \tag{1-30}$$

Here $n \equiv RC/\tau$ and $x \equiv t/\tau$. Equations (1-29) and (1-30) give the response when an exponential of rise time t_{r_1} ($= 2.2\tau$) is applied to a network of time constant RC (rise time $t_{r_2} = 2.2RC$). Thus $n \equiv RC/\tau \equiv t_{r_2}/t_{r_1}$. This response is shown in Figure 1-20 with a rise time of t_r. Actually the identical response also results when a step-function voltage is applied to a cascade of two networks of rise times t_{r_1} and t_{r_2}, provided that the second network does not load the first.

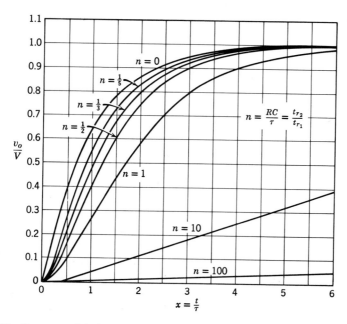

Fig. 1-20   Response of the low-pass RC network of rise time t_{r_1} ($= 2.2\,RC$) to an exponential of rise time t_{r_2} ($= 2.2\,\tau$); or identical response of two isolated cascaded low-pass RC networks to a step-function input. (From J. Millman and H. Taub, *Pulse, Digital, and Switching Waveforms*, McGraw-Hill, New York, 1965.)

A plot of t_r/t_{r_1} versus $t_{r_1}/t_{r_2} = 1/n$ is given in Figure 1-21. This plot indicates the relative rise time of two isolated low-pass RC networks in cascade with the resultant rise time t_r. This relative rise time also can be calculated by the following empirical equation, the error being 5 percent or less:

$$\frac{t_r}{t_{r_1}} = 1.05\sqrt{1 + n^2}. \tag{1-31}$$

Sweep (Ramp) Input

For a sweep input of the form $v_i = \alpha t$, the voltage output across the capacitor is given by

$$v_o = v_i - v_R = \alpha(t - RC) + \alpha RC\epsilon^{-t/RC}, \tag{1-32}$$

where v_R is the voltage across the resistor and is given by (1-17). If the time constant RC is small compared with the sweep duration T, the output will appear as in Figure 1-22a, where it is seen that the output follows the input but is delayed by one time constant RC from the input (except near the origin where there is distortion). The transmission error e_t is defined as the difference between input and output divided by the input at $t = T$. For $RC/T \ll 1$, we find

$$e_t \simeq \frac{RC}{T} = \frac{1}{2\pi f_2 T}. \tag{1-33}$$

If $RC/T \gg 1$, the output is very distorted, as in Figure 1-22b. Replacing the exponential in (1-32) by a series, we obtain

$$v_o \simeq \frac{\alpha t^2}{2RC}. \tag{1-34}$$

This quadratic response indicates that the circuit acts as an integrator.

RC Integrator

An integrator is a network in which the output signal is proportional to the mathematical integral of the input signal. One of the most commonly used forms is a low-pass RC network. If, as in Figure 1-16a, the time constant is very large compared with the time required for the input signal to make an appreciable change, the network is referred to as an integrator. In this case, the voltage drop across C will be very small relative to the drop across R, and we may consider that the total input v_i appears across R. Therefore the current is v_i/R and the output signal across C is

$$v_o = \frac{1}{C}\int i\,dt = \frac{1}{RC}\int v_i\,dt. \tag{1-35}$$

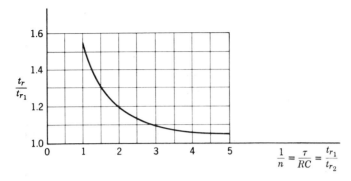

Fig. 1-21   Relative rise time of a cascade of two isolated low-pass *RC* networks. (If $t_{r_2} = 0$, then $1/n = \infty$ and $t_r/t_{r_1} = 1$.)

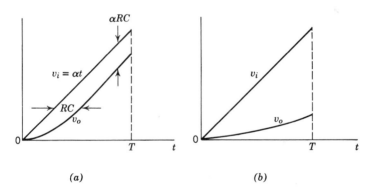

(a)   (b)

Fig. 1-22   Response of the low-pass *RC* network to a sweep input. (*a*) $RC/T \ll 1$; (*b*) $RC/T \gg 1$.

If $v_i = \alpha t$, the output is $\alpha t^2/2RC$, as given by (1-34). When time increases, the output will change from a quadratic to a linear function of time, as shown in Figure 1-22*a*. The integral of a constant is a linear function, and this agrees with the curves of Figure 1-19*d* which correspond to $RC/T \gg 1$.

A criterion for good integration in terms of steady-state analysis may be obtained by proceeding as in Sec. 1-3.

1-5   RLC CIRCUITS

RL Circuit

If the resistor *R* and capacitor *C* of the last two sections are substituted by a resistor R' and an inductor L, respectively, and if the time constant L/R' equals RC, then all the results remain unchanged.

The small air-core inductor or coil is used in small-time-constant applications. Now, let us see how a square wave is converted into pulses by means of a small time constant L/R. In the peaking circuit of Figure 1-23a, the bias voltage and the magnitude of the input are such that the triode operates linearly. In the equivalent circuit of Figure 1-23b, the output impedance R is the plate resistance r_p of the tube and the open-circuit voltage gain is the amplification factor μ of the triode. The peaking coil L acts as an open circuit at the time of an abrupt change in voltage. For a vacuum tube, with the output open-circuited, the change in plate voltage equals μ times the grid-voltage change. Therefore, as indicated in Figure 1-24, the peak of the output pulse equals μV, where V is the jump in voltage of the input signal. The output voltage falls or rises exponentially with a time constant L/R toward V_{pp}, the quiescent voltage. A peaking coil may also be used in the collector circuit of a transistor to obtain pulses.

Response of RLC Circuits

From the differential equations for the circuit of Figure 1-25, assuming a solution in the form ϵ^{st}, we find for the roots s of the characteristic equation[1]

$$s = -\frac{2\pi k}{T_0} \pm j\frac{2\pi}{T_0}(1-k^2)^{1/2}, \tag{1-36}$$

where k is the damping constant defined by $k \equiv (1/2R)(L/C)^{1/2}$, and T_0

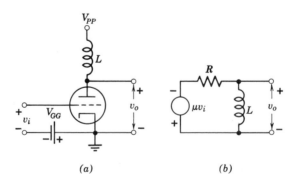

(a)   (b)

Fig. 1-23   (a) Peaking circuit using a peaking coil; (b) linear equivalent circuit.

[1] See M. G. Salvadori and R. J. Schwarz, *Differential Equations in Engineering Problems*, Prentice-Hall, Englewood Cliffs, N.J., 1954.

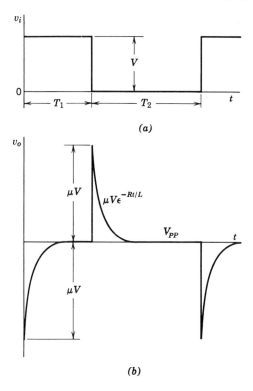

(a)

(b)

Fig. 1-24   Input v_i and output v_o for the circuit of Figure 1-23, with $R = r_p$. $L/R \ll T_1 < T_2$.

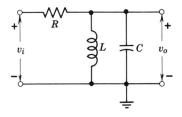

Fig. 1-25   An *RLC* circuit. (The response to a step-function input is given in Figure 1-26.)

is the resonant or undamped period defined by $T_o \equiv 2\pi (LC)^{1/2}$. If $k = 0$, then $s = \pm j2\pi/T_o$, and so the response is an undamped sine wave of period T_o. If $k = 1$, the two roots are equal, which corresponds to the critically damped case. If $k > 1$, there are no oscillations in the output, and the response is said to be overdamped. If $k < 1$, the output will be a sine wave whose amplitude delays with time, and the response is said to be underdamped. The parameter k is inversely proportional to the

$Q(\equiv \omega_o RC)$ of the circuit, which consists of a parallel combination of R, L, and C. Thus

$$Q \equiv \omega_o RC = \frac{2\pi RC}{T_o} = \frac{RC}{(LC)^{1/2}} = R\left(\frac{C}{L}\right)^{1/2} = \frac{1}{2k}.$$

If the input to the circuit of Figure 1-25 is a step voltage V and if the initial current through the inductor and the initial voltage across the capacitor are both zero, then the response is given by the following equations, in which $x \equiv t/T_o$:

1. For the case of critical damping, $k = 1$,

$$\frac{v_o}{V} = 4\pi x\,\epsilon^{-2\pi x} = \frac{4Rt}{L}\,\epsilon^{-2Rt/L}.$$

(1-37)

2. For the overdamped case[1], $k > 1$,

$$\frac{v_o}{V} \simeq \epsilon^{-\pi x/k} - \epsilon^{-4\pi kx}.$$

(1-38)

The second term is negligible compared with the first, except near $x = 0$. Thus (1-38) can be approximated by

$$\frac{v_o}{V} \simeq \epsilon^{-\pi x/k} = \epsilon^{-Rt/L}.$$

(1-39)

This result shows that the response approaches that for the zero-capacitance case (Figure 1-23) as k becomes much greater than unity.

3. For the underdamped case, $k < 1$,

$$\frac{v_o}{V} = (2k/\sqrt{1 - k^2})\,\epsilon^{-2\pi kx}\,\sin 2\pi\sqrt{1 - k^2}\,x.$$

(1-40)

The damped period is $T_o/(1 - k^2)^{1/2}$ and it is larger than the free period T_o.

Figure 1-26 gives the response of the circuit of Figure 1-25 for fixed L and C. The damping is varied by adjusting R. The curves for $k = 1$ and $k = 3$ are plotted from (1-37) and (1-38), respectively. The curves for $k < 1$ are given by (1-40). For $k = \infty$, $R = 0$ and $v_o/V = 1$. Note that if the damping factor k is adjusted to be somewhat less than unity, an excellent peaking circuit results. For fixed L and C the damping can be increased by shunting the LC circuit with an additional resistor.

---

[1] It is convenient to rewrite (1-36) as $s = -2\pi k/T_o \pm (2\pi k/T_o)(1 - 1/k^2)^{1/2}$. If we assume that k is large enough so that $4k^2 \gg 1$, then $(1 - 1/k^2)^{1/2} = 1 - 1/2\,k^2 - 1/8\,k^4 - \cdots \simeq 1 - 1/2\,k^2$, and we find for s the approximate values $-\pi/T_o\,k$ and $-4\pi k/T_o$. Subject to this restriction on the size of k, the response is given by (1-38).

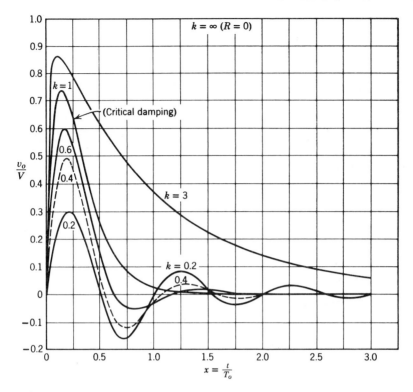

Fig. 1-26 Response of the circuit of Figure 1-25 for fixed L and C. $k \equiv (1/2\ R)\ (L/C)^{1/2}$. (From J. Millman and H. Taub, *Pulse, Digital, and Switching Waveforms*, McGraw-Hill, New York, 1965.)

Ringing Circuit

A ringing circuit is an *RLC* circuit which has nearly undamped oscillations. If the damping factor k is small, the circuit will ring for many cycles. Suppose the circuit would ring for N cycles before the amplitude decreases to $1/\epsilon$, or 0.37 of its initial value. Then, from (1-40) we see that this decrement results when $2\pi\ kx = 1$. Since $k = 1/2Q$ and $x = t/T_o = NT_o/T_o = N$, we have

$$Q = \pi N. \tag{1-41}$$

If the damping is small enough, the response approaches an undamped sine wave. The initial magnetic energy stored in the inductor is converted into electric energy in the capacitor at the end of one quarter cycle. Thus

$$\frac{LI^2}{2} = \frac{CV_{\max}^2}{2} \quad \text{or} \quad V_{\max} = \mathrm{I}\left(\frac{L}{C}\right)^{1/2}. \tag{1-42}$$

The following example illustrates the generation of a burst of oscillations by a switch circuit. Such a burst may be useful for timing purposes if its frequency is suitably chosen. The basic technique is shown in Figure 1-27. While the switch is closed, a current $I = V_{bb}/R_b$, corresponding to a stored energy $LI^2/2$, is maintained in the inductor. When the switch is opened, this current begins to charge the capacitor, initiating the alternating exchange of magnetic and electrostatic energy in the circuit. The peak capacitor voltage V_{max} is such that the energy $CV_{max}^2/2$ is the same (neglecting dissipation) as that initially in the inductor: $V_{max} = I(L/C)^{1/2}$. While the switch is open, the oscillations are damped by R_p, with $Q = R_p(C/L)^{1/2}$. When the switch is later closed, the damping due to R_b ($<< \sqrt{L/C}$) quickly brings the circuit back to its quiescent condition.

1-6   PULSE TRANSFORMERS

Introduction

Iron-cored transformers are used in steady-state ac circuits, in audio and rf circuits, and in the transmission and shaping of pulses which range in width from a fraction of a nanosecond to about 25 μsec. Figure 1-28 shows the schematic diagram of a transformer, including the source v_i and the load R_L. The primary inductance is L_p, the secondary inductance is L_s, and the mutual inductance is M. The coefficient of coupling K between primary and secondary inductance is defined by $K \equiv M/(L_pL_s)^{1/2}$. In a perfect transformer, $K = 1$, and

$$\frac{v_o}{v_i} = \frac{i_p}{i_s} = \sqrt{\frac{L_s}{L_p}} = \frac{N_s}{N_p} = \sqrt{\frac{R_L}{R_p}} = n, \qquad (1\text{-}43)$$

where v_o = output
$\quad\quad\quad v_i$ = input
$\quad\quad\quad i_p$ = primary current
$\quad\quad\quad i_s$ = secondary current
$\quad\quad\quad N_p$ = primary number of turns
$\quad\quad\quad N_s$ = secondary number of turns
$\quad\quad\quad n$ = transformation ratio
$R_p = R_L/n^2$ = effective load resistance reflected to primary side

A pulse transformer behaves as a reasonable approximation to an ideal transformer when used in connection with the fast waveforms it is intended to handle. The circuit of Figure 1-29 is equivalent to the circuit of Figure 1-28. In Figure 1-29, the resistance R_1 is the sum of the primary winding resistance and the generator impedance (assumed resistive); the

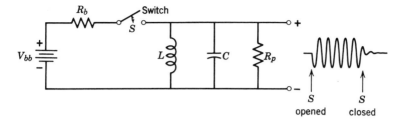

Fig. 1-27 Switch-excited *LCR* circuit (high *Q*) to generate burst of oscillations.

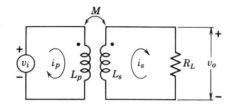

Fig. 1-28 Schematic diagram of a transformer circuit.

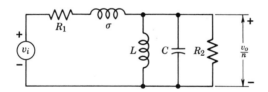

Fig. 1-29 Equivalent circuit of a transformer.

series inductance σ, called the leakage inductance, is the sum of the two series inductances; the shunt inductance L is the magnetizing inductance; the capacitance C is the total effective shunt capacitance (including $n^2 C_L$, the capacitance reflected from the secondary shunt-loading capacitance C_L); and the resistance R_2 is the combination of the load resistance R_L and the secondary winding resistance R'_2, so that $R_2 = (R_L + R_2')/n^2$. In a well-constructed pulse transformer the coefficient of coupling K differs from unity by less than 1 percent; the leakage inductance and the magnetizing inductance have the approximate values

$$\sigma \simeq 2L_p (1 - K) \tag{1-44}$$

and

$$L \simeq L_p. \tag{1-45}$$

For a simple transformer with a one-layer secondary wound over a one-layer primary (see Figure 1-30a), the total shunt capacitance is given by

$$C = (n^2 + n + 1) \frac{C_0}{3}, \qquad (1\text{-}46)$$

in which

$$C_0 \equiv \frac{\epsilon S \lambda}{d},$$

where ϵ = Dielectric constant[1] of medium separating windings
S = mean circumference of windings
λ = length of windings
d = separation between windings
For $n = 1$, $C = C_0$.

The Magnetizing and Leakage Inductances

The magnetizing inductance is the inductance presented at the input terminals when the transformer secondary is open-circuited. Similarly, the leakage inductance is the inductance presented at the terminals of the primary when the secondary is short-circuited.

For the simple magnetic circuit of Figure 1-30a, the primary inductance is given by

$$L_p = \mu A N_p^2 / l \qquad (1\text{-}47)$$

where l = mean length of magnetic path
A = cross-sectional area of core
N = number of primary turns
μ = magnetic permeability[2]

Figure 1-30a shows the geometrical arrangement of a simple transformer in which a single-layer secondary is wound over a one layer-primary. Consider this arrangement to see how the leakage inductance σ depends on the geometry. As the secondary is short-circuited, the output voltage is zero, and the net flux in the iron is zero. Thus $I_p N_p - I_s N_s = 0$, or $I_p N_p = I_s N_s$. Almost all the flux appears in the space between the windings, as shown in Figure 1-30b. These windings are considered as current sheets which are the same length λ (in the direction perpendicular to the current flow) as the coils are long. The value of the

---

[1] If the dielectric constant of the medium relative to free space is ϵ_r, then $\epsilon = \epsilon_r \epsilon_0$, where $\epsilon_0 = (36 \pi \times 19^9)^{-1} \, F/m$ is the permittivity of free space.
[2] If the permeability of the iron relative to free space is μ_r, then $\mu = \mu_r \mu_0$, where $\mu_0 = 4\pi \times 10^{-7} \, H/m$ is the permeability of free space.

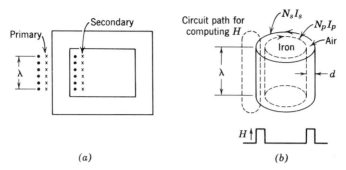

Fig. 1-30   (a) A single-layer primary is wound on a magnetic core, and a one-layer secondary is wound directly over the primary. A dot indicates current out of the page, and a cross indicates current into the page. (b) A schematic view of the windings considered as current sheets, and the magnetic-flux density between windings.

magnetic field intensity in the region between sheets is $H = I_p N_p/\lambda$. The energy density stored in the magnetic field is given by $\mu H^2/2$. Therefore the total energy stored is $W = \mu_0 H^2 V/2$, where V is the volume between windings, and where we have replaced μ by μ_0, the permeability of free space. The energy can also be calculated from $W = \sigma I_p^2/2$, since this magnetic energy (with the secondary shorted) may be considered to reside in the leakage inductance σ. Equating the above two expressions for W, we obtain

$$\sigma = \frac{\mu_0 H^2 V}{I_p^2} = \frac{\mu_0 N_p^2 V}{\lambda^2}, \qquad (1\text{-}48)$$

where all quantities are expressed in mks units. From this equation we see that σ is essentially independent of the magnetic circuit of the transformer, since the leakage flux is almost entirely in air. One of the main reasons for using high-permeability cores in pulse transformers is to have a large ratio of L_p/σ $(= \mu A \lambda^2/\mu_0 Vl)$. σ can be measured with a Q meter or an impedance bridge provided that the secondary is shorted. However, the effect of the resistance which is in series with σ should be taken into account. If the above experiment is repeated with the secondary open-circuited, then the magnetizing inductance L_p will be measured.

The Rise-time Response

The rise-time response near the front edge of the pulse is given by the high-frequency equivalent circuit shown in Figure 1-31, which is obtained from Figure 1-29 by neglecting the effect of L. The magnitude of the step-function input is V. From the differential equations for this circuit,

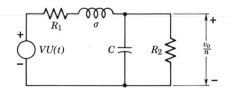

Fig. 1-31   High-frequency equivalent circuit used to calculate the rise-time response.

assuming a solution in the form ϵ^{st}, we find for the roots s of the characteristic equation

$$s = -\frac{2\pi}{T} k \pm j \frac{2\pi}{T} (1 - k^2)^{1/2}, \qquad (1\text{-}49)$$

where k is the damping constant defined by $k \equiv (R_1/\sigma + 1/R_2C)T/4\pi$, and T is the period defined by $T \equiv 2\pi (\sigma Ca)^{1/2}$, in which a is the amplification factor defined by $a \equiv R_2/(R_1 + R_2)$.

If $k = 0$, $s = \pm j2\pi/T$, and so the response is an undamped sine wave of period T. For the cases of $k = 1$, $k > 1$, and $k < 1$, the responses are given by the following equations in which $x \equiv t/T$ and $y \equiv v_0/naV$.

1. Critical damping, $k = 1$:

$$y = 1 - (1 + 2\pi x) \epsilon^{-2\pi x}. \qquad (1\text{-}50)$$

2. Overdamped, $k > 1$:

$$y = 1 - \frac{4k^2}{4k^2 - 1} \epsilon^{-\pi x/k} + \frac{1}{4k^2 - 1} \epsilon^{-4\pi kx}. \qquad (1\text{-}51)$$

If $4k^2 >> 1$,

$$y \simeq 1 - \epsilon^{-\pi x/k}. \qquad (1\text{-}52)$$

3. Underdamped, $k < 1$:

$$y = 1 - \left[\frac{k}{(1 - k^2)^{1/2}} \sin 2\pi (1 - k^2)^{1/2} x + \cos 2\pi (1 - k^2)^{1/2} x \right] \epsilon^{-2\pi kx}.$$

$$(1\text{-}53)$$

The positions x_m and magnitudes y_m of the maxima and minima[1] of the underdamped response are given by

$$x_m = \frac{m}{2(1 - k^2)^{1/2}} \qquad (1\text{-}54)$$

and

$$y_m = 1 - (-1)^m \epsilon^{-2\pi k x_m} \qquad (1\text{-}55)$$

[1] Equations (1-54) and (1-55) are derived by setting the derivative of (1-53) equal to zero.

where m is any integer. The maxima and minima are obtained for odd and even values of m, respectively.

The rise-time responses are plotted in Figure 1-32 for several values of k. We find from (1-50) or Figure 1-32 that for the critically damped case the rise time is

$$t_r = 0.53T = 3.35(\sigma Ca)^{1/2}. \qquad (1\text{-}56)$$

As defined previously, the rise-time t_r is the time interval required for the output to rise from 10 to 90 percent of its final value.

For many applications an overshoot in the output of 5 or 10 percent is acceptable. In such a case the overshoot will reduce the rise time (see Figure 1-32). Large step-up ratios (n) are seldom used in pulse transformers. Usually $n < 10$.

The Flat-top Response

The response during the flat top of the pulse is given by the low-frequency equivalent circuit of Figure 1-33a, which is obtained from Figure 1-29 by neglecting the effect of σ and C. Applying Thevenin's theorem, we obtain Figure 1-33b, where $a \equiv R_2/(R_1 + R_2)$ and $R \equiv R_1 R_2/(R_1 + R_2)$. The output is given by

$$y = \frac{v_0}{naV} = \epsilon^{-Rt/L}. \qquad (1\text{-}57)$$

If $Rt/L << 1$, then

$$y = \frac{v_0}{naV} \simeq 1 - \frac{Rt}{L}. \qquad (1\text{-}58)$$

Therefore the top of the output pulse will be tilted downward; the percent tilt is given by

$$P_{\text{tilt}} = (Rt_p/L) \times 100 \text{ percent}, \qquad (1\text{-}59)$$

where t_p is the pulse width.

The inductance L is constant so long as the iron does not begin to saturate. For a ferrite core the permeability is fairly constant for flux densities B up to a maximum B_m, which is of the order of 1500 to 5000 gauss (0.15 to 0.5 webers/m²). Saturation occurs if B exceeds the maximum value. $B = \phi/A$, where ϕ is the magnetic flux and A is the cross-sectional area of the core. Since $v_0 = N_s(d\phi/dt) = nN_pA(dB/dt)$, the flux density at the end of the pulse is given by

$$B = \int_0^{t_p} \frac{v_0}{nN_pA}\, dt = \frac{aVt_p}{N_pA}, \qquad (1\text{-}60)$$

provided that the top of the pulse is flat and equals naV. When the input

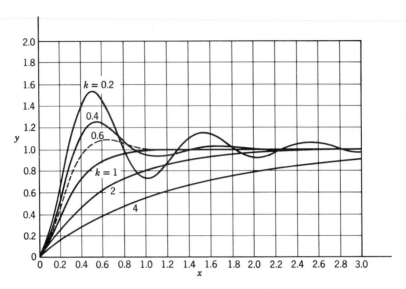

Fig. 1-32   The rise-time response of a pulse transformer is given by the circuit of Figure 1-31. $y \equiv v_0/naV$ and $x \equiv t/T$. (From J. Millman and H. Taub, *Pulse, Digital, and Switching Waveforms*, McGraw-Hill, New York, 1965.)

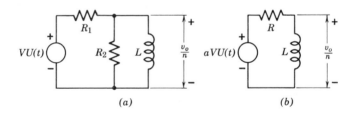

Fig. 1-33   (a) Low-frequency equivalent circuit used to calculate the flat-top response of a pulse transformer; (b) Thevenin equivalent for the circuit of (a). $a \equiv R_2/(R_1 + R_2)$ and $R \equiv R_1 R_2/(R_1 + R_2)$.

duration exceeds the value of t_p given by (1-60) with $B = B_m$, the output will drop rapidly. Notice that it is the volt-second product which determines the maximum flux density. In any particular application we must be sure not to saturate the core.

To minimize both tilt and backswing we require, as evident from (1-59), that $L >> Rt_p$. Accordingly, if, say, $t_p = 0.1$ μsec and $R = 200$ ohms, $Rt_p = 20$ μH, and a magnetizing inductance $L = 1$ mH ($= 50 \times 20$ μH) is, for all practical purposes, large enough.

The Complete Pulse Response

The composite waveform $y(t)$ is obtained by first plotting the exponential portions at the top of the pulse which are given by (1-57). Then the positive- and negative-peak overshoots, given by $y_m - 1$ of (1-55), are superimposed on the exponential.

The response beyond the pulse width $(t > t_p)$ is obtained as follows. A pulse can be considered to be the sum of a step of voltage $+V$ whose discontinuity occurs at $t = 0$ and a step of voltage $-V$ whose discontinuity occurs at $t = t_p$. Therefore, if the transformer response to a step V at $t = 0$ is $y(t)$, then the output for $t > t_p$ is $y(t) - y(t - t_p)$. For the flat-top response $y(t) = \epsilon^{-Rt/L}$, and hence

$$y(t) - y(t - t_p) = \epsilon^{-Rt/L} - \epsilon^{-R(t - t_p)L}$$

$$= (1 - \epsilon^{Rt_p/L}) \, \epsilon^{-Rt/L} \qquad (1\text{-}61)$$

for $t > t_p$. For the underdamped case $(k < 1)$ the trailing edge of the output waveform will contain the same high-frequency oscillations as are present on the leading edge, and the area of the part of the waveform above the zero axis will equal the area which is below the zero axis. The high-frequency oscillations can be reduced to zero by increasing the loading on the transformer (or by decreasing the amplification factor a).

Example 1-1

The parameters of a pulse transformer are $L = 5$ mH, $\sigma = 40 \ \mu$H, $C = 50$ pF, $R_1 = 200$ ohms, $R_2 = 2$ K and $n = 1$. Calculate and plot the complete response to a 2-μsec 1.0-volt pulse.

Solution   (a) The rise-time response is calculated as follows:

$$a = \frac{R_2}{R_1 + R_2} = \frac{2000}{200 + 2000} = 0.909,$$

$$T = 2\pi(\sigma C a)^{1/2} = 2\pi(40 \times 10^{-6} \times 50 \times 10^{-12} \times 0.909)^{1/2} = 0.267 \ \mu\text{sec},$$

$$k = \left(\frac{R_1}{\sigma} + \frac{1}{R_2 C} \right) \frac{T}{4\pi}$$

$$= \left(\frac{200}{40 \times 10^{-6}} + \frac{1}{2 \times 10^3 \times 50 \times 10^{-12}} \right) \frac{2.67 \times 10^{-7}}{4\pi}$$

$$= 0.318.$$

For $k < 1$, y is given by (1-53); that is

$$y = 1 - \left[\frac{k}{(1 - k^2)^{1/2}} \sin 2\pi \, (1 - k^2)^{1/2} \, \frac{t}{T} + \cos 2\pi \, (1 - k^2)^{1/2} \, \frac{t}{T} \right] \epsilon^{-2\pi kt/T}$$

Substituting numerical values into the above equation, we obtain

$$y = 1 - (0.325 \sin 22.3t + \cos 22.3t)\, \epsilon^{-7.48t},$$

where t is expressed in μsec.

From (1-54) we find that the maxima and minima occur at

$$t_m = \frac{mT}{2(1-k^2)^{1/2}} = 0.141m,$$

where $m = 1, 2, 3, \ldots$; and from (1-55) we find that the magnitudes at t_m are

$$y_m = 1 - (-1)^m\, \epsilon^{-1.01m}.$$

(b) The flat-top response is given by (1-57); that is,

$$y = \epsilon^{-Rt/L} = \epsilon^{-0.0364t} \simeq 1 - 0.0364t$$

where t is expressed in μsec. The percentage tilt of the top of the pulse is $3.64t_p = 7.28$ percent.

(c) For $t > t_p$ the response is given by (1-61), which for the given value of transformer parameters reduces to

$$y = (1 - \epsilon^{0.0728})\, \epsilon^{-0.0364t} = -0.0758\, \epsilon^{-0.0364t}.$$

At $t = t_p += 2\ \mu$sec, $y = -0.071$.

(d) The complete response (up to $t = 6\ \mu$sec) is plotted in Figure 1-34.

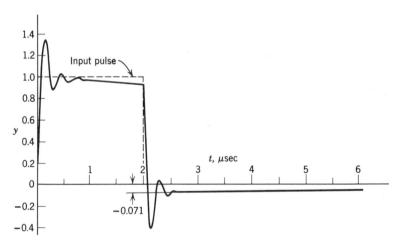

Fig. 1-34 Complete pulse response of the transformer whose parameters are given in Example 1-1. $v_0 = naVy = 0.909\ y$.

The long undershoot for $t > t_p$ will slowly approach the zero axis, so that the net area under the curve will equal zero.

Winding-and-core General Considerations

In a small pulse transformer, the preservation of the pulse shape is more important than efficiency of operation. The winding resistances may therefore be quite large, often as large as 10 percent of the load or generator resistances. When the load and generator resistances are high, a large series-leakage inductance may be much more readily tolerated than a large shunt capacitance. In this case the windings may be spaced far apart. If the load and generator resistances are very small, a close spacing may be preferred.

The smallest core on which there is space available to place the windings is normally selected, and it is a good idea to check, with the aid of (1-60), to be sure that the core is not saturated at the peak of the pulse.

There is, in principle at least, one method that will both minimize σ and C and increase L_p arbitrarily. This method involves employing a core material whose magnetic permeability is infinite. Since in such a case the required turns are minimal, we may shrink the spacing between "windings" so that σ vanishes without introducing an appreciable capacitance. Pulse transformers employ as core material such alloys as Hipersil (Westinghouse) [μ_r (max) $\simeq$ 12,000] or Permalloy (Western Electric) [μ_r (max) $\simeq$ 80,000] or ferrites (see below).

Because of the eddy currents, the permeability actually achieved in pulse transformers is very much less than the maximum values indicated above. The effective permeability of Hipersil is of the order of 400 for microsecond pulses. A pulse-transformer core is often formed by winding a continuous strip of thin high-permeability alloy, as shown in Figure 1-35.

Cores molded from a magnetic ceramic such as sintered manganese-zinc ferrite that are excellent for pulse transformers are now available. The maximum permeability of this material is not very great, but its resistivity is at least 10^7 times that of Hipersil or Permalloy. Because of this high resistivity the skin effect due to eddy currents is very small and an effective permeability of the order of 1000 is attained. Also the core loss is very small, and a Q of the order of 5 to 15 is obtained at a frequency of 1 MHz. One form in which ferrite cores are commercially available is shown in Figure 1-36a. An end view of the complete core, assembled by putting two halves together, is shown in Figure 1-36b. The primary inductance of a core whose dimensions are given in Figure 1-36 may be calculated to be $L_p = 1.1 \, N_p^2 \, \mu$H, to within 10-percent accuracy.

Ferrite cores are also commercially available in the form of toroids (doughnuts) in very small sizes; these make excellent pulse transformers for nanosecond applications.

Fig. 1-35   A pulse-transformer core is formed by winding a continuous trip of thin high-permeability alloy.

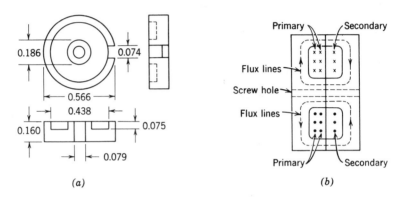

(a)   (b)

Fig. 1-36   (a) Three views of a small ferrite pot (or cup) core. Dimensions are in inches. (Courtesy of Ferroxcube Corporation.) (b) An end view of the transformer assembled by putting two halves together.

1-7   DELAY LINES

Introductory Considerations

Delay lines are passive four-terminal networks which have the property that a signal impressed at the input terminals appears at the output terminals at the end of a delay time t_d. Delay times ranging from a few nanoseconds to hundreds of microseconds are obtainable with electromagnetic lines. The applications of delay lines are numerous. For example, a cathode-ray oscilloscope which is to be used for observing fast waveforms has a built-in delay line so that the input signal which also triggers the sweep is delayed slightly before being applied to the vertical-deflection circuit. If the sweep was not allowed to start before the signal was applied, the first portion of the waveform might not be visible on the scope face. Other applications of delay lines occur in distributed amplifiers, in precise time measurement, and in digital-computer systems.

A uniform lossless transmission line, terminated in its characteristic impedance Z_0, can be used as a delay line. If C and L are the capacitance and inductance, respectively, of a unit length of the line, then the characteristic impedance (resistive) is given by $Z_0 = (L/C)^{1/2}$. For an ideal transmission line, an arbitrary waveform impressed on the input terminals will appear without distortion at the output terminals after a delay time t_d. If the line is terminated in the characteristic impedance Z_0, no reflection will take place when the signal reaches the end of the line.

Figure 1-37 shows three types of transmission lines, together with the expressions for the corresponding capacitance C per unit length. These lines are the coaxial cable, the single wire over ground, and the two parallel wires. For these three types of line the relationships among L, C, Z_0, and u are expressed as

$$L = \frac{1}{u^2 C} \quad \text{and} \quad Z_0 = \frac{1}{uC}, \tag{1-62}$$

where L is the inductance per unit length and u is the velocity of propagation. For a uniform line, $u = (\mu \epsilon)^{-1/2}$, where μ and ϵ are the magnetic permeability and the permittivity, respectively, of the medium between the conductors of the line. The capacitance C, inductance L, and impedance Z_0 are extremely insensitive to changes in spacing.

For a lossless coaxial line

$$Z_0 = 138 \, \epsilon_r^{-1/2} \log \frac{D}{d},$$

where ϵ_r is the dielectric constant of the material between the inner and outer conductor. When attenuation in the line results principally from ohmic losses in the conductors, the loss (for a fixed D) is a minimum for $D/d = 3.6$. For this ratio and for $\epsilon_r = 2.3$ (for polystyrene, polyethylene, or Teflon), $Z_0 = 51$ ohms. Most commercially available coaxial lines

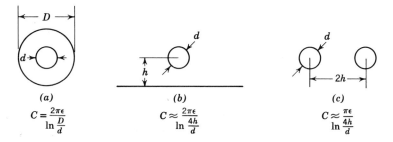

$$(a) \qquad C = \frac{2\pi\epsilon}{\ln \frac{D}{d}}$$

$$(b) \qquad C \approx \frac{2\pi\epsilon}{\ln \frac{4h}{d}}$$

$$(c) \qquad C \approx \frac{\pi\epsilon}{\ln \frac{4h}{d}}$$

Fig. 1-37 Three types of transmission lines, together with expressions for their capacitance per meter: (a) coaxial cable; (b) wire over ground; (c) parallel wires.

have impedances from 50 to about 200 ohms. For a wire-over-ground line, $Z_0 = 138 \log (4\ h/d)$; and if $h = d$, $Z_0 \simeq 83$ ohms. A widely used two-wire line is the 300-ohm line used as an antenna lead-in for television receivers.

The time delay per unit length T is given by[1]

$$T = \sqrt{LC} = CZ_0 = \frac{L}{Z_0}. \tag{1-63}$$

Taking a nominal value $Z_0 = 100$ ohms, we have, for an air dielectric, $T \simeq 1$ nsec/ft $\simeq 3.3$ nsec/m, $C \simeq 10$ pF/ft $\simeq 0.33$ pF/cm, and $L \simeq 100$ nH/ft $\simeq 3.3$ nH/cm.

Delay-line Pulse Shaping

The termination of the delay line is matched if the load equals the characteristic impedance $Z_0 = R_0$. Also, the input end of the line is matched if the impedance of the signal source equals Z_0. Mismatching a delay line at one end causes delayed reflections to appear at the other. These can be used for pulse shaping. Their outstanding feature is that the profile of the reflection is similar to that of leading edge of the input, within the limitations of the rise time of the line itself. Use of a delay line can transform the initial step into a neat rectangular pulse. This is called delay-line clipping or delay-line differentiation. The pulse width produced in a delay line is dependent only on the line delay T, a relatively constant and reliable parameter.

Two main types of mismatch for the line are a short circuit and an open circuit. The first returns a reflection with inverted voltage, wiping out the voltage existing on the line (for a step-function input) and carrying back the news that the line is really a short circuit. The second inverts the current in the reflection; this cancels the existing line current and carries the news of the open-circuit termination back to the input. In either case the input end of the line must be properly matched if further reflections are to be avoided.

If source and load can be idealized (impedance much smaller or much larger than $R_0 = Z_0$), the clipping circuits shown in Figure 1-38 can be assembled. The shorted lines are used in shunt positions and the open ones are used in series. In every case the line looks like an impedance R_0 until the reflection has had time to return, after time $2T$. The output pulse amplitudes are therefore half those of the input, as indicated, and the clipped pulse widths are $2T$. When source or load has finite imped-ance, this fact must be taken into account for matching the line. Resistive

---

[1]For a uniform lossless line the delay per meter T is also given by $T = (\mu\epsilon)^{1/2} = 1/\mu$. For air $T = (3 \times 10^8)^{-1}$ sec/m $= 3.3$ nsec/m. For a medium of relative dielectric constant ϵ_r, the delay is $3.3\ \epsilon_r^{1/2}$ nsec/m.

impedances may often be incorporated as part of the matching resistor shown in Figure 1-38.

Delay-line Differentiation

A delay line can be used in several different ways to convert a step function into a square wave of controlled duration. One way is to impress the signal on a short-circuited delay line from a source whose impedance matches the characteristic impedance of the delay line (Figure 1-39). After a time interval equal to twice the propagation time of the line, the signal returns from the short-circuited end to cancel the signal at the input end. The accuracy of the cancellation depends upon the accuracy of the impedance match and the phase distortion and attenuation in the delay line. If care is taken to choose the terminating network that will compensate for the inevitable lumped capacitance at the driving end and that will compensate, at least partially, for the phase distortion, the residual ringing signals observed with commercially available helically wound delay cables can be successfully reduced to a few percent of the signal size. The "rear porch" that results from attenuation in the line (Figure 1-40*a*) can be effectively eliminated by passing the signal through an appropriate *RC* differentiating circuit. However, the top of the resultant signal (Figure 1-40*b*) is no longer perfectly flat.

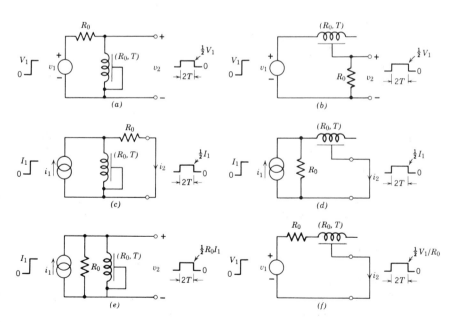

Fig. 1-38 Delay-line clipping circuits with idealized source and load impedances. Delay-line characteristic impedance $Z_0 = R_0$.

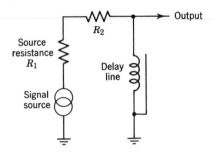

Fig. 1-39 Delay-line differentiating circuit. R_2 is chosen so that $R_1 + R_2$ is equal to the delay-line impedance.

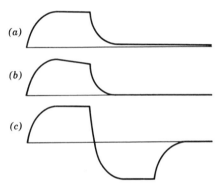

Fig. 1-40 Single-delay-line differentiated signal: (*a*) showing "rear porch" due to delay-line attenuation and (*b*) with rear porch removed by *RC* differentiation. (*c*) Double-delay-line-differentiated signal (bipolar pulse).

There is one way of eliminating the rear porch without resorting to additional differentiation so that the output signal will retain its flat top. The technique involves the use of a delay line terminated in its characteristic impedance at the receiving end and, preferably but not necessarily, at the sending end. The delayed signal is subtracted from the input signal in a difference amplifier. The input signal may be slightly attenuated at the difference-amplifier input to compensate for the attenuation of the delayed signal in the delay line. Such an arrangement is shown in Figure 1-41. The delay line is terminated at the receiving end by R_4, and at the sending end by the parallel combination of $(R_1 + R_2)$ and R_3. The tap on R_3 is adjusted to compensate for attenuation in the line. Since the delay line is terminated at both ends, much poorer matches can be tolerated without producing excessive ringing.

Applying delay-line clipping at two places in the amplifier results in a bipolar signal (see Figure 1-40c) in which the positive and negative areas are equal. The base-line[1] location is independent of counting rate, and both pileup overloading and low-frequency noise can be minimized. However, with double-delay-line differentiation (clipping) the midband noise is increased by an additional factor of $\sqrt{3}$ with respect to that obtained with single-delay-line clipping and by an overall factor of $\sqrt{6}$ with respect to single or double RC differentiation.

1-8   INTRODUCTION TO DIODES

The Thermionic Diode

The thermionic diode is the simplest form of electron tube. It contains two electrodes, a cathode, and an anode (plate). A diode can conduct current in only one direction. If it is used as an element in a circuit, any portion of the input voltage which makes the anode negative with respect to the cathode will stop the conduction of current through the tube. If a sinusoidal voltage is applied to a diode, through a cathode series resistor, then the current through the resistor will flow during only half of each cycle. This is known as half-wave rectification. When the anode is positive with respect to cathode, the diode resistance is low, and it is called forward resistance. When the anode is negative, the diode resistance is much higher, and it is called back resistance. The sum of the

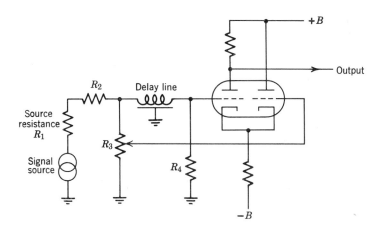

Fig. 1-41   Alternative delay-line differentiating circuit. The delayed signal is subtracted from the input signal in the difference amplifier. (See Sec. 1-15.)

[1] The base line is the datum from which pulse heights are measured.

voltage drop across two resistors in series equals the source voltage. Thus, if the back resistance is very high, almost all the voltage drop will appear across the tube and almost none will appear across the resistor. The source voltage is almost all across the load resistance during conduction of the tube. A thermionic diode has a much higher back resistance than does a semiconductor diode. One of the most important ratings of the tube is the maximum-peak inverse-anode (plate) voltage. Thermionic diodes would be chosen in pulse circuits principally for their high reverse voltage rating and their very low reverse current; semiconductor diodes are considered standard in most circumstances.

The Semiconductor Diode

The semiconductor diode is the simplest type of semiconductor device. It is basically a *p-n* junction similar to those shown in Figure 1-42. The *n*-type material which serves as the negative electrode is called the cathode, and the *p*-type material which serves as the positive electrode is called the anode. The arrow symbol used for the anode represents the direction of conventional current flow; electron current flows in a direction opposite to the arrow. When the positive terminal of an external battery is connected to the *n*-type material and the negative terminal to the *p*-type material, as in Figure 1-42*a*, the space charge region at the junction becomes effectively wider, and the potential gradient (represented by the imaginary battery) increases until it approaches the potential of the external battery. Current flow is then extremely small because no voltage difference (electric field) exists across either the *p*-type or the *n*-type region. Under these conditions, the *p-n* junction is said to be reverse biased. When the positive terminal of the external battery is connected to the *p*-type material and the negative terminal to the *n*-type material, as shown in Figure 1-42*b*, the space-charge

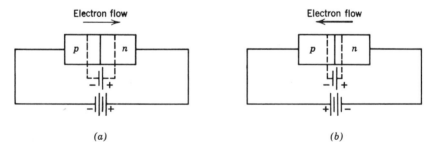

(a)   (b)

Fig. 1-42   Electron current flow in biased *p-n* junction: (*a*) reverse bias; (*b*) forward bias.

region becomes effectively narrower, and the energy barrier (formed by the potential gradient) decreases to an insignificant value. Excess electrons from the *n*-type material can then penetrate the space-charge region, flow across the junction, and move by way of the holes in the *p*-type material toward the positive terminal of the battery. This electron flow continues so long as the external voltage is applied. Under these conditions the junction is said to be forward biased.

For an ideal *p-n* junction diode the current *I* is related to the voltage *V* by the equation

$$I = I_0 \, (\epsilon^{V/nV_T} - 1), \qquad\qquad (1\text{-}64)$$

where I_0* is the junction scale current, nV_T is the junction scale voltage, and *n* equals 1 for germanium and approximately 2 for silicon. The symbol V_T stands for the electron-volt equivalent of temperature and is given by $V_T \equiv kT/e$, in which *k* is the Boltzmann constant ($k = 1.38 \times 10^{-23}$ joule/°K), *e* is the electronic charge ($e = 1.602 \times 10^{-19} C$), and *T* is the absolute temperature. Substituting, we find that $V_T = (T/11,600)$ V and that at room temperature ($T = 300°$K) $V_T = 26$ mV. Thus the scale voltage (nV_T) is 26 to 50 mV for *p-n*–junction diodes at room temperature (26 mV for Ge and 40 to 50 mV for Si).

Equation (1-64) is plotted, together with the actual characteristics of a silicon junction diode, in Figure 1-43. In the forward direction, the exponential increase in current merges into a linear section as soon as the purely ohmic series elements in the diode structure become important. In the back direction the reverse current increases very rapidly as soon as a certain reverse voltage is exceeded. This is the breakdown region. The voltage at which breakdown occurs is called the avalanche or Zener voltage. One of the most important ratings of semiconductor diodes is the maximum peak reverse voltage (PRV), that is, the highest amount of reverse voltage which can be applied to a specific diode before the avalanche breakdown point is reached. PRV ratings range from about 50 volts to as high as 1000 volts for some single-junction diodes.

Rectifiers and Voltage Multipliers

Figure 1-44*a* shows a circuit in which two rectifiers are used as a full-wave rectifier. In this case the alternating current is more efficiently converted to pulsating direct current. The frequency of the output from the full-wave rectifier is twice that from the half-wave rectifier, and it is therefore easier to filter (convert to smooth direct current with little or no

---

*When the diode is reverse-biased and $|V|$ is several times V_T, $I \simeq -I_0$. For this reason, I_0 is often referred to as the reverse saturation current.

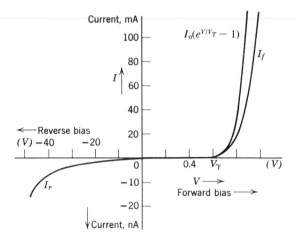

Figure 1-43   Characteristic of a silicon junction diode (type 1N3605). An exponential form has been fitted to the I_f curve to show deviations at the higher currents; the ideal reverse-conduction curve ($I_r = -I_0$) coincides with the voltage axis on the scale of the diagram. (Courtesy of General Electric Company.)

ripple). It is necessary to use a center-tapped transformer to supply the ac input wave to the two diodes in Figure 1-44a. (It is also necessary to use a center-tapped transformer to supply the filaments if two rectifier tubes are used instead of the two semiconductor diodes.) One-half of the transformer and one diode operate at any given instant. When the input signal is such that the anode of D_1 is positive, the anode of D_2 is negative. At this instant D_1 will conduct and D_2 will not. When D_2 is positive with respect to its cathode, it will conduct. The result is that half of each cycle is contributed by each diode. The voltage output can be doubled by using four diodes instead of two. This is called a full-wave bridge rectifier, as shown in Figure 1-44b.

A voltage-doubler circuit of simple form is shown in Figure 1-45a. The circuit derives its name from the fact that its dc voltage output can be as high as twice the peak value of ac input. Basically, a voltage doubler is a rectifier circuit arranged so that the output voltages of two half-wave rectifiers are in series. The action of a voltage doubler may be described briefly as follows. On the positive half-cycle of the ac input, that is, when the upper side of the ac input line is positive with respect to the lower side, the upper diode passes current and feeds a positive charge into the upper capacitor. As positive charge accumulates on the upper plate of the capacitor, a positive voltage builds up across the capacitor. On the next half-cycle of the ac input, when the upper side of the line is negative with respect to the lower side, the lower diode passes current so that a positive voltage builds up across the lower capacitor. So long as no current is drawn at the output terminals from the capacitor, each capacitor can charge up to the same magnitude of voltage as the peak value of the ac input. Thus the total voltage across the capacitors is as high as twice the peak value of ac input. The arrangement shown in Figure 1-45*a* is called a full-wave voltage doubler because each rectifier passes current to the load on each half of the ac input cycle. A voltage tripler is shown in Figure 1-45*b*, and a voltage quadrupler is shown in Figure 1-45*c*.

Zener Diode

An avalanche or Zener diode is a special type of semiconductor diode. It is designed with adequate power-dissipation capabilities to operate in the breakdown region, as shown in Figure 1-46*a*, hence it can be employed as a voltage-reference or constant-voltage source. Such a diode is used characteristically in the manner indicated in Figure 1-46*b*, where it is substituting for the gaseous glow tube conventionally employed in this circuit. The source V and resistor R are chosen so that the diode is initially operating in the breakdown region. The diode will regulate the load voltage V_Z against variations in load current and against variations in supply voltage V, since in the breakdown region large changes in diode current produce only small changes in diode voltage.

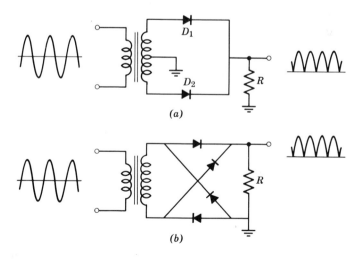

(a)

(b)

Fig. 1-44 Full-wave rectification of sinusoidal signals. (*a*) Center-tapped transformer; (*b*) bridge circuit.

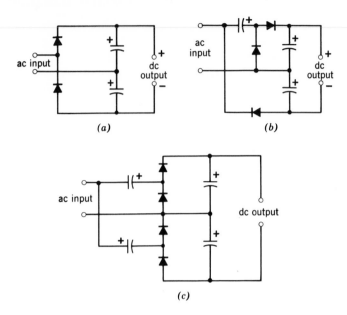

Fig. 1-45 (*a*) Voltage doubler; (*b*) voltage tripler; (*c*) voltage quadrupler.

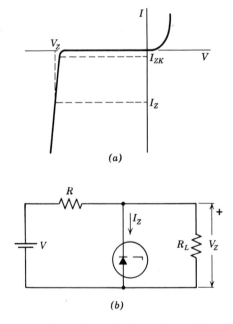

Fig. 1-46 (*a*) Volt-ampere characteristic of a Zener diode; (*b*) a circuit in which a Zener diode is used as a source of reference potential V_Z.

The diode will continue to regulate until the circuit operation requires the diode current to fall to I_{ZK}, which is in the neighborhood of the knee of the diode volt-ampere curve. The upper limit on diode current is determined by the power-dissipation rating of the diode.

In the mechanism of diode breakdown the thermally generated electrons and holes acquire sufficient energy from the applied potential to produce new carriers by removing valence electrons from their bonds. These new carriers, in turn, produce additional carriers again through the process of disrupting bonds. This cumulative process is referred to as avalanche multiplication. It results in the flow of large reverse currents, and the diode finds itself in the region of avalanche or Zener breakdown. Even if the initially available carriers do not acquire sufficient energy to disrupt bonds, it is possible to initiate breakdown through a direct rupture of the bonds because of the existence of the strong electric field.

Zener diodes are available with voltages as low as about 2 volts. Below this voltage we may use diodes in the forward direction for reference and regulating purposes. This is possible because the forward characteristic (see Figure 1-43) is something like the reverse characteristic. When a high-voltage reference is required it is usually advantageous to use two or more Zener diodes in series rather than a single diode. This combination will allow higher voltage, higher dissipation, lower temperature coefficient, and lower dynamic resistance. Silicon diodes operated in avalanche breakdown are available with maintaining voltages from several to several hundred volts and with power ratings up to 50 watts.

The Tunnel Diode

In a *p-n* junction diode, the depletion layer or space charge region constitutes a potential barrier at the junction. This potential barrier restrains the flow of majority carriers from one side of the junction to the other. The width of the depletion layer (junction barrier) varies inversely as the square root of the impurity concentration. A *p-n* junction diode of the conventional type has an impurity concentration of about 1 part in 10^8. With this amount of doping the width of the depletion layer is of the order of 5 μ (5×10^{-4} cm). If the concentration of impurity is greatly increased, say to 1 part in 10^3, then the width of the junction barrier is reduced from 5 μ to about 100 Å (10^{-6} cm). This thickness is only about one fiftieth the wavelength of visible light. For a barrier as thin as this, quantum mechanics indicates that there is a great probability that an electron will penetrate through the barrier. The quantum-mechanical behavior is known as "tunneling," and therefore such a high-impurity-density p-n–junction device is called the "tunnel diode."

With the tunnel-diode characteristic of Figure 1-47, for small reverse voltages the reverse current is high. Also, for small forward voltages (up to 50 mV for Ge) the resistance remains small (of the order of 5 Ω). At the peak current I_p corresponding to the voltage V_p the slope dI/dV of the characteristic is zero. If V is increased beyond V_p, then the current decreases, and so the dynamic conductance $g = dI/dV$ is negative. Thus the tunnel diode exhibits a negative-resistance characteristic between the peak current I_p and the valley current I_v (corresponding to the valley voltage V_v). At V_v the conductance is again zero, and beyond this point the resistance becomes and remains positive. At the peak forward voltage V_F the current again reaches the value I_p. The explanation for the tunneling phenomenon is found in quantum mechanics. In order for an electron in the n region to cross the junction and enter a hole in the p region, it must have an energy which falls within a certain specified band of energies. Electrons with other energy levels are forbidden to enter these holes. Hence, as the forward voltage is increased, the energy of the electrons in the n region increases, first into a region which allows the electrons to cross the barrier and enter holes causing the current to increase, and then into a forbidden region causing the current to decrease; after this the current again begins to climb. Here the tunneling effect is no longer necessary, and current increases with increasing forward bias until the avalanche voltage is exceeded. Thus, the portion of the characteristic beyond V_v is caused by the injection current in an ordinary p-n-junction diode. The remainder of the characteristic is a

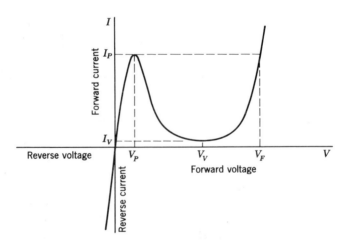

Fig. 1-47   Volt-ampere characteristic of a tunnel diode.

result of the tunneling phenomenon in the highly doped diode. Since the current between the peak and the valley is decreasing with increasing voltage, the resistance of the tunnel diode is said to be negative. For currents whose values are between I_v and I_p the curve is triple-valued because each current can be obtained at three different applied voltages. It is this multivalued feature which makes the tunnel diode useful in pulse and digital circuitry.

Figure 1-48a gives the tunnel diode symbols. Figure 1-48b shows the small-signal model in the negative-resistance region. Here $-R_n$ is the negative resistance, R_s is ohmic resistance, L_s depends upon the lead length and the geometry of the diode package, and C is the junction capacitance which depends upon the bias and is usually measured at the valley point. For $I_p = 10$ mA, the typical values of these parameters are $-R_n = -30\Omega$, $R_s = 1\Omega$, $L_s = 5$nH, and $C = 20$pF. Table 1-1 gives the important static characteristics of various tunnel diodes. Our principal interest in the tunnel diode is its application as a very high-speed switch.

Table 1-1
Typical Tunnel Diode Parameters

	Ge	GaAs	Si
I_p/I_V	8	15	3.5
V_p, V	0.055	0.15	0.065
V_V, V	0.35	0.50	0.42
V_F, V	0.50	1.10	0.70

1-9 BASIC DIODE CIRCUITS

The Diode Rules

There are many different types of diode circuits in addition to the rectifiers described previously. These circuits can be understood more easily by observing the following diode rules.

The first diode rule is that for practical diodes (nonlinear devices), the applied signal must be sufficiently large to exploit the nonlinearity with telling effect. Very small signals will swing the diode over only a short segment of the characteristic, which then appears almost linear. The required voltage level may be found from (1-64), remembering that the scale voltage V_T is 25 to 50 mV for junction diodes at room temperature.

Another rule follows from the trivial observation that diodes conduct in one direction only. This implies that any current which follows must

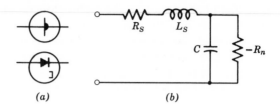

Fig. 1-48   (*a*) Tunnel diode symbols; (*b*) small-signal model.

have a dc component, and consequently that a dc return path must be provided around every diode.

Note that a forward diode may be considered to be a short circuit, and a reverse diode open, as in the ideal case. Remember that in order for a diode to conduct it must be forward-biased by at least the cut-in (or break-point) voltage V_γ [approximately 0.2 V for germanium or 0.6 V for silicon (see Figure 1-43)].

Clipping Circuits

Clipping circuits are also known as voltage (or current) limiters, amplitude selectors, or slicers. They are used to choose for transmission that part of an arbitrary waveform which lies above or below some particular reference voltage level. They normally fall into one of the following configurations: (a) a series combination of a diode, resistor, and reference supply (the diode may be the input circuit of a vacuum tube triode, pentode, or transistor); (b) a network consisting of several diodes, resistors, and reference voltages; (c) two emitter-coupled or cathode-coupled triodes operating as an overdriven difference amplifier.

A common clipping circuit of Figure 1-49 illustrates how the diode limits the signal on the top or on the bottom and how the double-diode clipper limits at two independent levels. The input signal v_i is a sine wave. The reference levels V_{R1} and V_{R2} are equal. With both switches S_1 and S_2 open, the output signal is a sine wave (Figure 1-49*b*). Suppose v_i is larger than V_{R1} or V_{R2}. Then, with S_1 open and S_2 closed, the diode D_2 is forward-biased on each negative half-cycle when the anode is positive with respect to the cathode, and hence the resulting waveform at the output appears as in Figure 1-49*c*. With S_2 open and S_1 closed, the resulting waveform at the output appears as in Figure 1-49*d*. With S_1 and S_2 both closed, the diode clippers are used in pair, and the resulting waveform at the output appears as in Figure 1-49*e*.

Clamping Circuits

INTRODUCTION   To create a function with a periodic waveform it is necessary to establish the recurrent positive or negative extremity at some constant reference level V_R. The circuits used to perform this function are known as clamping circuits, since they restrain the extremity of the waveform from going beyond V_R in the steady state. The need to establish the extremity of the positive or negative signal excursion at V_R often appears in connection with a signal which has passed

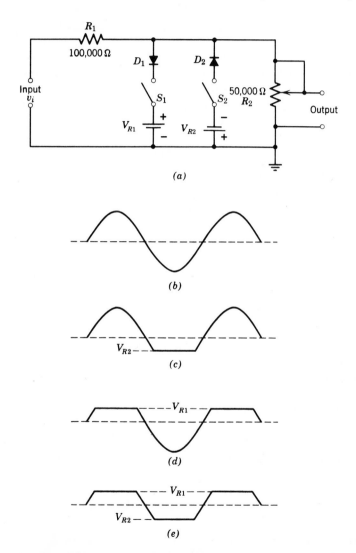

Fig. 1-49 (a) Common clipping circuit; (b) sine-wave output signal with S_1 and S_2 both open; (c) positive-peak output signal with S_1 open and S_2 closed; (d) negative-peak output signal with S_1 closed and S_2 open; (e) clipped output signal with S_1 and S_2 both closed.

through a capacitive coupling network. Such a signal has lost its dc component, and the clamping circuit introduces a dc component. For this reason the clamping circuit is often called a dc restorer, dc reinserter, or dc inserter. Note that the dc component so introduced is not identical with the dc component lost in transmission.

Figure 1-50 shows a clamping circuit in which the resistance R_s of the signal source v_s is taken into account. During conduction the diode is assumed to be a forward resistance R_f. The resistance R_f will lie in the range of tens to hundreds of ohms, depending on the type of diode used. The break-point or cut-in voltage V_γ in the diode characteristic is about 0.2 V for Ge and 0.6 V for Si (see Figure 1-43). Beyond V_γ the forward current rises very rapidly. Now, for simplicity, we assume that the diode breakpoint V_γ occurs at zero voltage. The precision of operation of the circuit depends on the condition that $R >> R_f$. The equivalent circuit in Figure 1-51a applies when the diode is conducting and that in Figure 1-51b applies when the diode is not conducting.

THE TRANSIENT APPROACH TO THE STEADY STATE   In order to see how the transient waveform approaches the steady state, we shall make use of the equivalent circuits of Figure 1-51 and proceed as in Example 1-2.

Example 1-2

In the circuit of Figure 1-50, $R_s = R_f = 100\ \Omega$, $R = 10$ K, and $C = 1.0\ \mu$F. At t = 0 there is applied a symmetrical square-wave signal v_s of amplitude 10 V and frequency 5 kHz, as indicated in Figure 1-52. Compute and draw the first several cycles of the output waveform.

Solution   Assume that the capacitor C is initially uncharged. At $t = 0$, the diode is forward-biased, and the equivalent circuit of Figure 1-51a is applicable. Thus at the first 10-V jump of the input v_s, the output v_o jumps to $+5$ V. The capacitor now charges so that v_o decays toward zero exponentially, with a time constant

$$\tau = (R_s + R_f)C = 200\ \mu\text{sec.}$$

Since the frequency is 5 kHz, the period $T = (5000)^{-1}$ sec $= 200\ \mu$sec; then at $t = T/2$, the output v_o, as indicated in Figure 1-52, has fallen to

$$v_0 \left(T = \frac{T}{2} \right) = 5\ \epsilon^{-T/2\tau} = 5\ \epsilon^{-1/2} = 5 \times 0.606 = 3.0\ \text{V}$$

At this time, the voltage across R_f or the voltage across R_s is 3.0 V, and so the capacitor voltage v_A is 4.0 V. At $t = (T/2)+$, the input drops back to zero and the diode does not conduct; we now use the equivalent circuit of Figure 1-51b. In this circuit, $v_A = 4.0$ V and $v_s = 0$, so that, neglecting R_s compared with R, $v_0 = -4.0$ V, as in Figure 1-52. The output now again starts to decay toward zero. However, the time constant is now $RC = (10\text{K})(1.0 \ \mu\text{F}) = 10,000 \ \mu\text{sec}$, which is 100 times larger than the time $T/2 = 100 \ \mu\text{sec}$. Hence the decay is negligible.

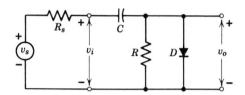

Fig. 1-50 A clamping circuit with the source resistance R_s.

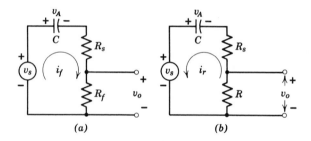

(a)   (b)

Fig. 1-51 Circuits equivalent to the circuit of Figure 1-50 (a) when the diode is conducting, (b) when the diode is not conducting. $R_r \gg R \gg R_f$.

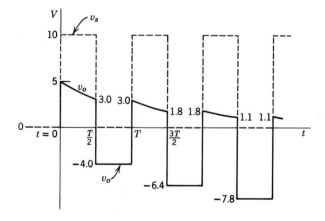

Fig. 1-52 Example of the transient waveform in Figure 1-50.

Because in the interval $t = T/2$ to $t = T$ the capacitor voltage has not changed ($v_A = 4.0\,\text{V}$), then at $t = T+$, the diode is forward-biased, and the output returns to $+3.0\,\text{V}$ ($= 10 - 4 - 3\,\text{V}$). Again the output decays toward zero exponentially with $\tau = 200\ \mu\text{sec}$, and the output at $t = 3T/2$ has fallen to $3\ \epsilon^{-1/2}$ or $1.8\,\text{V}$. The remaining calculations are repetitions of those above, and the results are shown in Figure 1-52. Cycle by cycle the output waveform is approaching the steady-state case, at which the positive excursion of the waveform is clamped approximately to zero.

The calculation of the clamped level is according to the clamping-circuit theorem. This theorem states that for any input waveform the ratio of the area under the output-voltage curve in the forward direction to that in the reverse direction (A_f/A_r) is equal to the ratio R_f/R, or

$$\frac{A_f}{A_r} = \frac{R_f}{R}. \tag{1-65}$$

If this theorem is applied to the waveform in Figure 1-52, it is found that the peak of v_0 is at $0.1\,V$.

THE STEADY-STATE OUTPUT WAVEFORM   Now consider the square wave of Figure 1-53a applied to the clamping circuit of Figure 1-50. The general form of the output shown in Figure 1-53b is determined by the four voltages V_1, V_1', V_2, and V_2'. At $t = 0-$ when $v_s = V''$ and $v_0 = V_2'$, the equivalent circuit of Fig. 1-51b is applicable, and the capacitor voltage is

$$v_A = V'' - \frac{R + R_s}{R} V_2'. \tag{1-66}$$

At $t = 0+$ the input signal jumps to V' and the output jumps to V_1; Figure 1-51a is applicable, and the capacitor voltage is

$$v_A = v_s - \frac{R_f + R_s}{R_f} v_0 = V' - \frac{R_f + R_s}{R_f} V_1. \tag{1-67}$$

Since the capacitor voltage cannot change instantaneously, (1-67) equals (1-66), or

$$V'' - \frac{R + R_s}{R} V_2' = V' - \frac{R_f + R_s}{R_f} V_1. \tag{1-68}$$

Since $V' - V'' = V$, (1-68) becomes

$$V = \frac{R_f + R_s}{R_f} V_1 - \frac{R + R_s}{R} V_2'. \tag{1-69}$$

Equation (1-69) is obtained by considering conditions at $t = 0-$ and $0+$.

Similarly, by considering conditions at $t = T_1-$ and T_1+, we obtain

$$V = \frac{R_f + R_s}{R_f} V'_1 - \frac{R + R_s}{R} V_2. \qquad (1\text{-}70)$$

During the interval T_1 the diode is conducting. Hence

$$V'_1 = V_1 \, \epsilon^{-T_1/(R_f + R_s)C}. \qquad (1\text{-}71)$$

Likewise, during the interval T_2 the diode is not conducting. Hence

$$V'_2 = V_2 \, \epsilon^{-T_2/(R + R_s)C}. \qquad (1\text{-}72)$$

Subtracting (1-70) from (1-69), we obtain

$$\frac{R_f + R_s}{R_f} (V_1 - V'_1) - \frac{R + R_s}{R} (V'_2 - V_2) = 0$$

or

$$\Delta_f = \frac{R_f}{R_f + R_s} \frac{R + R_s}{R} \Delta_r, \qquad (1\text{-}73)$$

where

$$\Delta_f \equiv V_1 - V'_1, \; \Delta_r \equiv V'_2 - V_2. \qquad (1\text{-}74)$$

Since R_s is usually much smaller than R, the tilt Δ_f in the forward direction is almost always less than the tilt Δ_r in the reverse direction. Only when $R_s << R_f$ are the two tilts almost equal.

PRACTICAL CLAMPING CIRCUITS   In order to maintain perfect flatness of the positive and negative peaks of a square wave in a practical situation, it will normally be required that $(R_f + R_s)C << T_1$ and $(R + R_s)C >> T_2$. A square wave or pulse waveform, after restoration, typically appears as in Figure 1-53c. During the interval T_2 there is a small tilt Δ_r, and at the beginning of the interval T_1 a sharp spike of magnitude Δ_f appears. The capacitor recharges through the diode in a very short time, and during the remainder of the time T_1 no appreciable diode current flows. From (1-73) the overshoot Δ_f will normally be smaller than the tilt Δ_r. From (1-70) to (1-74), the voltages are found to be

$$V'_1 = 0, \qquad V_2 = -\frac{R}{R + R_s} V, \qquad (1\text{-}75)$$

$$V'_2 = V_2 \, \epsilon^{-T_2/(R + R_s)C}, \qquad (1\text{-}76)$$

$$V_1 = \Delta_f = \frac{R_f}{R_f + R_s} \frac{R + R_s}{R} (V'_2 - V_2). \qquad (1\text{-}77)$$

From the above equations we can see that even if we assume C arbitrarily large, unless R_s is zero the distortion in the output signal is inevitable.

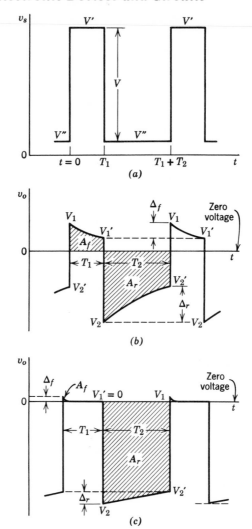

Fig. 1-53   (a) A square wave applied to the clamping circuit of Figure 1-50; (b) the general form of the output; (c) the expected form of the output of the clamping circuit with a square-wave input for $(R_f + R_s) C \ll T_1$ and $(R + R_s) C \gg T_2$.

This distortion, however, is not readily apparent in the case of a square wave, which already has a flat top.

   If the diode in Figure 1-50 is reversed, it can be shown (Example 1-3) that the negative rather than the positive extremity of the signal will be established at zero. If the circuit is modified to include a fixed voltage V_R, as in Figure 1-54a, the positive extremity (or negative extremity, if the diode is reversed) of the output will be established at V_R.

Sometimes the device which receives the output signal of the restorer of Figure 1-54a bridges a resistance across the output, and it is necessary to operate the circuit as in Figure 1-54b. This arrangement will operate perfectly well, provided that the amplitude of the input signal is adequately large. If the diode is to conduct, so that clamping may take place, the positive excursions of the signal at X must at least equal V_R.

Since the diode break point actually occurs at a voltage V_γ, the basic clamping circuit of Figure 1-55 is equivalent to that of Figure 1-54b with $V_R = V_\gamma$. Thus, if a silicon diode ($V_\gamma \approx 0.6\,\mathrm{V}$) is used, the circuit will function properly only if the positive excursion of the signal above its average value exceeds 0.6 V. If an external reference voltage V_R is added to the diode, then the clamping level is $V_R + V_\gamma$.

The clamping-circuit theorem can be generalized to

$$\frac{A_f - (V_R + V\gamma)\,T_1}{A_r} = \frac{R_f}{R},\qquad (1\text{-}78)$$

where T_1 is the interval over which the diode is forward-biased, and $R \gg R_f$.

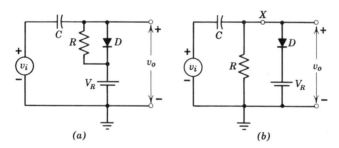

(a)   (b)

Fig. 1-54 (a) The dc restorer clamps to the reference voltage V_R; (b) a modification of the circuit of (a).

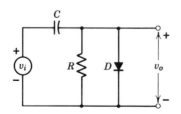

Fig. 1-55 A basic clamping circuit.

Example 1-3

(a) A square wave with $T_1 = 1$ msec and $T_2 = 1$ μsec has an amplitude of 10 V. This signal is applied to the clamping circuit of Figure 1-50, in which $R_s = 0$, $R_f = 50$ Ω, $R = 50$ K, and $V_\gamma = 0$. Assume that the capacitor C is arbitrarily large, so that the output is a square wave without tilt. Find where, on the waveform, the zero level is located. (b) If the waveform is inverted so that $T_1 = 1$ μsec and $T_2 = 1$ msec, locate the zero level. (c) If the diode is inverted but the input is as in (b), locate the zero level. (d) If in (a), (b), and (c) a silicon diode ($V_\gamma \simeq 0.6$ V) is used, find the clamping levels.

Solution

(a) The output waveform and zero level are shown in Figure 1-56a. Since $R_s = 0$, the peak-to-peak amplitude of the output signal is the same as that for the input signal, namely 10 V. We have $A_f = 1000\ V_1$ μsec volt and $A_r = 10 - V_1$ μsec volt. From (1-65) we obtain

$$\frac{A_f}{A_r} = \frac{1000\ V_1}{10 - V_1} = \frac{R_f}{R} = \frac{50}{50 \times 10^3} = 10^{-3},$$

and we find that $V_1 = 10^{-5}$ V. This indicates that only one millionth of the input waveform is above the zero level. Therefore clamping of the broad base line of the waveform is quite precise.

(b) In Figure 1-56b, $A_f = V_1$ and $A_r = (1000)(10 - V_1)$. From (1-65) we have

$$\frac{V_1}{1000\ (10 - V_1)} = 10^{-3},$$

or $V_1 = 5$ V. The zero level is now not near the positive peak but is halfway down the waveform, and the circuit has done very poorly as a clamp.

(c) Positive voltages now reverse bias the diode and negative voltages forward bias the diode, since the diode has been inverted. Comparing Figures 1-56c and a we see that one is inverted with respect to the other, and therefore $V_1 = 10^{-5}$ V, as in (a).

(d) An application of (1-78) yields the following results. In (a) and (c), $V_1 = 0.6 + 10^{-5} \simeq 0.6$ V, so that the broad base line is now clamped to $V_\gamma = 0.6$ V rather than to ground. In (b), $V_1 = 5.3$ V instead of 5.0 V.

An Example of a Half-wave Rectifier Used as a Power Supply

Figure 1-57 shows a typical half-wave rectifier circuit. For sufficient rating of peak reverse voltage, the six diodes are used in series. Also, for the required high-voltage reference, the sixteen voltage-reference tubes are connected in series. Each of these tubes is shunted by a 470-K resistor so that a voltage divider is formed. The voltage rating of the filter capacitors C_1 and C_2 must be adequate. In high-voltage power-supply circuits, low-leakage capacitors and low-leakage transformers are essential. The resistors R_1 and R_2 and capacitors C_1 and C_2 are used as a filter network, which smooths out the ripple of the diode output and increases rectifier efficiency. The limiting resistor R_3 and the dc input of the series voltage-reference tubes are selected so that initially the tubes are operating in the region of Townsend discharge (analogous to the avalanche breakdown in a Zener diode).

The diode of type 1N3563 (RCA) is a diffused-junction silicon rectifier. Its maximum peak reverse voltage is 1000 volts; its maximum rating of forward current (average) is 0.3 amp at 75° C. The miniature voltage-reference tube OG3/85A2 (Amperex) is one of cold cathode discharge types, and is for use in stable regulated power supplies and similar applications. For its typical operation, a dc operating current of 6 mA is recommended; the dc operating voltage at 6 mA (variation from tube to

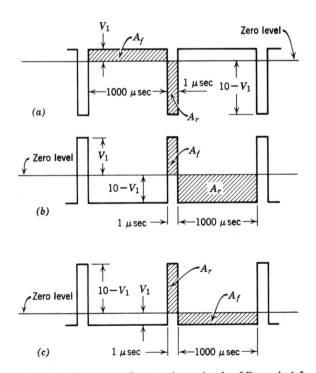

Fig. 1-56 Output waveforms and zero levels of Example 1-3.

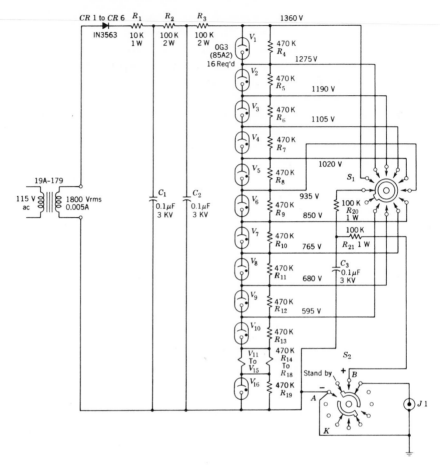

Fig. 1-57 A typical half-wave rectifier circuit—high voltage power supply. (From the manual *Operating and Maintenance Instructions for 400 Channel Analyzer System, Model 20631/2,* Radiation Counter Laboratories, Inc.)

tube) is 83 to 87 V; the ac resistance at 6 mA (average value) is 300 Ω; and the temperature coefficient of operating voltage is -2.7 mV/°C. The maximum ratings of max dc starting voltage, max dc operating current, and min dc operating current are 125 volts, 10 mA, and 1 mA, respectively.

1-10 CHARACTERISTICS OF VACUUM TRIODE TUBES

When a third electrode, called the grid, is placed between the cathode and plate (anode), the tube is referred to as a triode. The grid usually consists of relatively fine wire, wound on two support rods (side rods), that extends the length of the cathode. The purpose of the grid is to control the flow of plate current. When a tube is used as an amplifier, a negative dc voltage is usually applied to the grid. Under this condition the grid does not draw appreciable current.

There are many different types of tubes used for various purposes. The typical triodes 5963 (or 12AU7) and 5965 were designed for use in high-speed digital computers. The characteristic curves for 5965 are given in Figure 1-58. If the region near small plate voltages is ignored, then the positive-grid curves are very similar in shape and spacing to those for negative-grid values. Therefore, if the grid signal is supplied from a source of low impedance, so that the loading effect on the source due to the flow of grid current may be ignored, the tube will continue to operate linearly even if the grid signal makes an excursion into the positive-grid region. This linearity will continue so long as the grid current is a small fraction of the total cathode current. The variation of the tube parameters with plate current is given in Figure 1-59. These parameters are the amplification factor μ, the dynamic plate resistance r_p, and the transconductance g_m. The amplification factor μ is the ratio of the

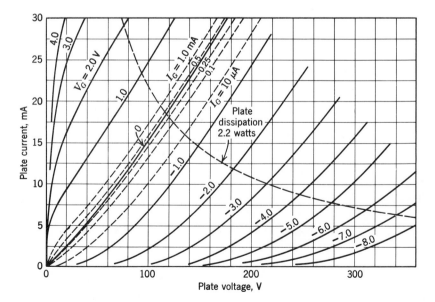

Fig. 1-58   Grid-plate characteristics of 5965 tube. The dashed lines are loci of constant grid current. $I_G = 400\ \mu A$ at $V_G = 0V$; $I_G = 10\ \mu A$ at $V_G = -0.5V$.

change in plate voltage dV_p to a change in grid voltage dV_G in the opposite direction, subject to the condition that the plate current I_P remain constant. The plate resistance r_p is the resistance of the path between cathode and plate to the flow of alternating current; it is the ratio of a small change in plate voltage dV_P to the corresponding change in plate current dI_p, subject to the condition that the grid voltage V_G remain constant. The transconductance g_m is a factor which combines in one term the amplification factor and the plate resistance; it is the quotient of the first divided by the second. By analogy with the definition of the dynamic plate resistance, the dynamic grid resistance r_g is given by dV_G/dI_G, where V_G and I_G are the instantaneous values of grid voltage and current, respectively. The static grid resistance r_G is defined as the ratio V_G/I_G. From the grid volt-ampere characteristics of Figure 1-60 it appears that the difference in values between the static and dynamic resistances is not great, except possibly for small grid voltages. From Figure 1-60 we find that for a 5965 tube, 250 ohms is a reasonable value for r_G.

From Figure 1-58 we estimate that the grid current is $I_G = 0.25$ μA at $V_G = -1$ V. If a grid leak resistor (R_g) of 4 MΩ is connected from grid to cathode of the 5965 tube, then a negative bias of 1 V will be developed. Hence the grid current at zero grid-to-cathode voltage, and even for slightly negative grid voltages, is often large enough to have an appreciable effect on the operation of a circuit.

1-11   TRANSISTORS

Basic Transistor Circuits

When a second junction is added to a semiconductor diode to provide power or voltage amplification, the resulting device is known as a transistor. The three regions of the device are called the emitter, the base, and the collector, as shown in Figure 1-61a. In normal operation, the emitter-to-base junction is biased in the forward direction, and the collector-to-base junction in the reverse direction. Different symbols shown in Figures 1-61b and 1-61c are used for n-p-n and p-n-p transistors to show the difference in the direction of current flow in the two types of devices. The emitter for both types is identified by an arrow indicating forward conduction.

There are three basic ways of connecting transistors in a circuit: common base (CB), common emitter (CE), and common collector (CC) (emitter follower). They are analogous to the three basic triode connections: common grid, common cathode, and common plate (cathode follower), respectively. In the CB (or grounded-base) configuration of Figure 1-62, the signal is applied to the emitter-base circuit and extracted

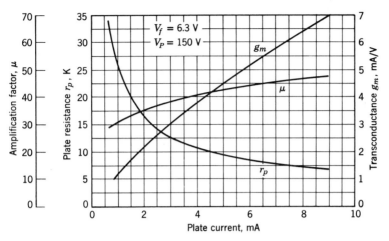

Fig. 1-59   Average small-signal parameters of 5965 tube.

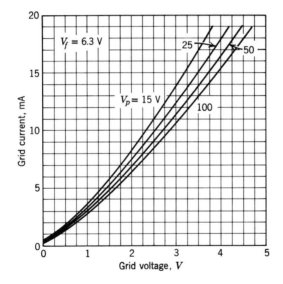

Fig. 1-60   Average volt-ampere grid characteristics of the 5965 tube.

from the collector-base circuit. Since the input or emitter-base circuit has a low impedance, on the order of 0.5 to 50 ohms, and the output or collector-base circuit has a high impedance, on the order of 1000 ohms to 1 MΩ, the voltage or power gain in this configuration may be on the order of 1500. The direction of the arrows in Figure 1-62 indicates electron

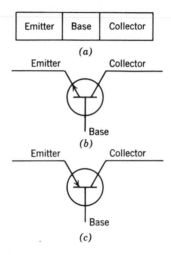

Fig. 1-61   Fundamental diagram and schematic symbols for transistors. (a) Functional diagram; (b) n-p-n transistor; (c) p-n-p transistor.

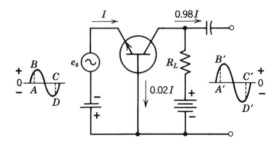

Fig. 1-62   Common-base (CB) circuit configuration (or connection). (From RCA *Transistor Manual*, 1967.)

current flow. Most of the current from the emitter flows to the collector; the remainder flows through the base. In practical transistors, from 95 to 99.5 percent of the emitter current reaches the collector. The current gain of the *CB* circuit, therefore, is always less than unity (usually on the order of 0.95 to 0.995). As shown by the waveforms in Figure 1-62, there is no voltage phase reversal between the input and the output of a *CB* amplifier.

In the *CE* (or grounded-emitter) configuration of Figure 1-63, the signal is applied to the base-emitter circuit and extracted from the collector-emitter circuit. The input (base-emitter) impedance is about 20 to 5000 Ω, and the output (collector-emitter) impedance is about 50 to 50,000 Ω. The power gain, of the order of 10,000, can be realized with a

CE circuit because it provides both current gain and voltage gain. The typical current gain is about 50. The input signal voltage undergoes a phase reversal of 180°, as shown by the waveforms in Figure 1-63.

In the *CC* (or grounded-collector) connection or emitter follower of Figure 1-64, the signal is applied to the base-collector circuit and extracted from the emitter-collector circuit. Since the input impedance of the transistor is high and the output impedance low in this configuration, the voltage gain is less than unity and power gain is normally lower than that obtained in either a *CB* or a *CE* circuit. The emitter follower is used primarily as an impedance-matching device. There is no phase reversal of the signal between the input and the output.

Transistor Characteristics

ALPHA (α) AND BETA (β) CURRENT-TRANSFER RATIOS. Vacuum tubes are voltage controlled; the current through the tube is controlled by the voltage applied to the grid. On the other hand, transistors are current-

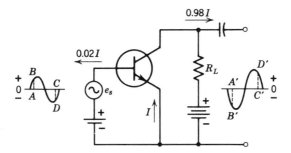

Fig. 1-63 Common-emitter (CE) circuit configuration (or connection). (From RCA *Transistor Manual*, 1967.)

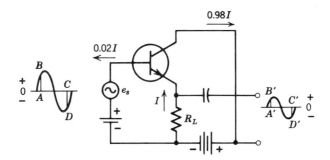

Fig. 1-64 Common-collector (CC) circuit configuration (or connection). (From RCA *Transistor Manual*, 1967.)

controlled devices; an input current controls an output current, and the ratio of the controlled output current to the controlling input current is called the forward current-transfer ratio. The common-base forward current-transfer ratio is called alpha (α) or h_{fb}. It is the ratio of the change in collector current to the corresponding change in emitter current; for ac signals, $\alpha = i_c/i_e$. In the *CB* circuit of Figure 1-62, the dc alpha or α_{FB} is the ratio of the dc collector current I_C to the dc emitter current I_E:

$$\alpha_{FB} = \frac{I_C}{I_E} = \frac{0.98I}{I} = 0.98.$$

The common-emitter forward current-transfer ratio is called beta (β) or h_{fe}. It is the ratio of the change in collector current to the corresponding change in base current; for ac signals, $\beta = i_c/i_b$. In the *CE* circuit of Figure 1-63, the dc beta or h_{FE} is

$$h_{FE} = \frac{I_C}{I_B} = \frac{0.98I}{0.02I} = 49.$$

Since $i_b = i_e - i_c$, and since $\beta = i_c/i_b$ and $\alpha = i_c/i_e$, it can be shown that

$$\beta = \frac{\alpha}{1 - \alpha}, \qquad \text{and} \qquad \alpha = \frac{\beta}{1 + \beta}.$$

Figure 1-65 indicates typical electrode currents in a *CE* circuit under no-signal conditions and with a 1-μA signal applied to the base. The signal current of 1 μA in the base causes a change of $147 - 98$ or 49 μA in the collector current. Thus $\beta = i_c/i_b = 49$.

FREQUENCY CUTOFF.   The frequency cutoff of a transistor is defined as the frequency at which the value of alpha or beta drops to 0.707 times its 1-kHz (or kc/sec) value. The gain-bandwidth product is the frequency at which the common-emitter forward current-transfer ratio (β) is equal to unity. These characteristics provide an approximate indication of the useful frequency range of the device and help to determine the most suitable circuit configuration for a particular application.

EXTRINSIC TRANSCONDUCTANCE.   The extrinsic transconductance can be defined as the quotient of a small change in collector current divided by the small change in emitter-to-base voltage producing it, subject to the condition that other voltages remain unchanged.

CUTOFF CURRENTS.   Cutoff currents are small dc reverse currents which flow when a transistor is biased into nonconduction. They consist of leakage currents, which are related to the surface characteristics of

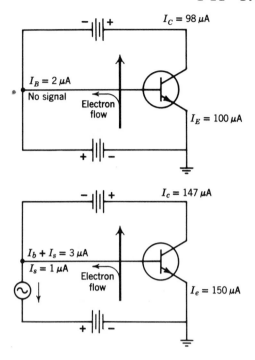

Fig. 1-65   Typical electrode currents under no-signal and signal conditions.

the semiconductor material, and saturation currents, which are related to the impurity concentration in the material and which increase with increasing temperatures. Collector-cutoff current (I_{CBO}) is the dc current which flows in the reverse-biased collector-to-base circuit when the emitter-to-base circuit is open. Emitter-cutoff current (I_{EBO}) is the dc current which flows in the reverse-biased emitter-to-base circuit when the collector-to-base circuit is open.

TRANSISTOR RATINGS.   The ratings of a transistor limit operating conditions to prevent permanent damage by voltage breakdown or over-heating. The conditions that must be specified are the maximum values of collector reverse voltage, emitter reverse voltage, collector current, internal-power level and collector-junction temperature.

BREAKDOWN VOLTAGES.   Transistor breakdown voltages define the voltage values between two specified electrodes at which the crystal structure changes and current begins to rise rapidly. The voltage then remains relatively constant over a wide range of electrode currents.

Breakdown voltages can be measured with the third electrode open, shorted, or biased in either the forward or reverse direction. The symbols BV_{CEO}, BV_{CER}, BV_{CES}, and BV_{CEV} are sometimes used to designate collector-to-emitter breakdown voltages with the base open, with external base-to-emitter resistance, with the base shorted to the emitter, and with a reverse base-to-emitter voltage, respectively.

THREE REGIONS OF OPERATION. When turned fully on, the transistor is said to be operating in the saturation region. When turned fully off, it is said to be operating in the cutoff region. The region between is often called the transition or active region. The transistor may be used as a switch by simply biasing the base-emitter diode junction off (cutoff) or on (saturation). Adjusting base-emitter bias to some point approximately midway between cutoff and saturation will place the transistor in the active, or linear, region of operation. When operating within this region, the transistor is able to amplify. Figure 1-66 locates the cutoff, active,

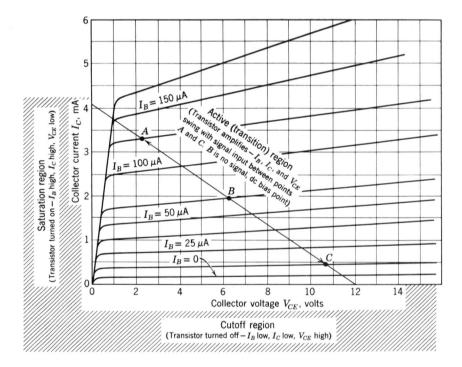

Fig. 1-66 Three regions of transistor operation on a typical set of common-emitter collector characteristic curves. The points A, B, C are on the load line. (From *Transistor Manual*, 7th ed., General Electric Company, 1964.)

and saturation regions of transistor operation on a typical set of collector characteristic curves. For amplifier operation, base-emitter dc bias will be approximately 0.2 V for germanium and 0.6 V for silicon (see Table 1-5).

Field-effect Transistors

A field-effect transistor (FET) is essentially a semiconductor current path whose conductance is controlled by applying an electric field perpendicular to the current. The electric field results from reverse biasing a *p-n* junction. Thus in the FET of Figure 1-67, the current flow from source to drain is controlled by the voltage (field) applied to the gate. Here the action is analogous to the vacuum tube, in which plate current is controlled by grid voltage; in the field-effect transistor, drain current is controlled by gate voltage. Like the vacuum tube, the FET has high input impedance. Another advantage is low inherent noise level.

There are two types of FET[1], *N* channel and *P* channel. The former is made of *N*-type semiconductor material, on which *P*-type regions (gates) are deposited (Figure 1-67). In the *P*-channel FET, *N*-type gates are deposited on *P*-type material. Unlike the junction transistor, the input (gate) is reverse-biased, so no current flows in the gate, and the current from source to drain then depends on the voltage applied to the gate. Figure 1-68 shows a way to bias an FET—self-bias by source resistor R_s. The low noise (cold) FET has been satisfactorily used as a first stage in the preamplifier of a semiconductor radiation spectrometer.

1-12 LOW-FREQUENCY SMALL-SIGNAL TRANSISTOR MODEL

The transistor, like the vacuum tube, is a nonlinear active device. However, if the bias is chosen so that it moves the transistor's operation to the more linear portion of its characteristics, and if the bias is such that the largest ac signal to be amplified is small compared to the dc bias current and voltage, then the imperfect linearity is acceptable, resulting in amplification with low signal distortion. The restriction to small signal levels will lead to more accurate equivalent circuits composed of linear circuit elements and internal linear generators. This allows the analyst the use of conventional linear-circuit analysis.

The transistor is a three-terminal device. One of these terminals will be used as common, the other two as input and output. In view of the

[1] The FET's are also divided into the following two types: (a) junction field-effect transistor (abbreviated JFET, or FET); (b) insulated-gate field-effect transistor (IGFET), more commonly called the metal-oxide-semiconductor (MOS) transistor (MOST or MOSFET).

wealth of information available on the analysis of four-terminal devices, however, it is more convenient to analyze the transistor as a four-terminal device. Figure 1-69 illustrates that the three-terminal network is just a special case of the four-terminal network in which the common terminal serves in both the input and output portions of the circuit.

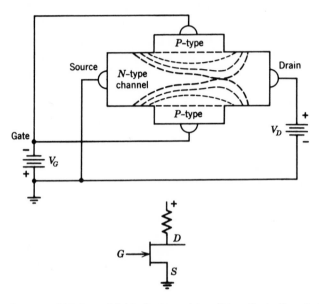

Fig. 1-67   Elements of N-channel field-effect transistor. P-type "gates" control the resistance between N-type source and N-type drain regions. (From Transistor Reference Handbook, M. W. Lads Company, Philadelphia, Pennsylvania.)

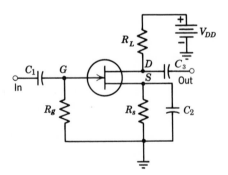

Fig. 1-68   A way to bias an FET—self-bias by resistor in source circuit.

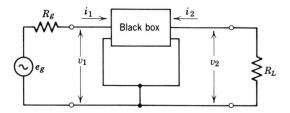

Fig. 1-69 Four-terminal linear network representation of the transistor. The common terminal (base) serves in both the input and output portions of the network.

Open-Circuit Impedance Parameters

The input and output voltages (v_1 and v_2) of a black box (Figure 1-69) are expressed in terms of the input and output currents (i_1 and i_2) and are given by the generalized equations

$$v_1 = i_1 z_{11} + i_2 z_{12} \qquad (1\text{-}79a)$$

and

$$v_2 = i_1 z_{21} + i_2 z_{22}, \qquad (1\text{-}79b)$$

where z_{11}, z_{12}, z_{21}, and z_{22} are the input, reverse transfer, forward transfer, and output impedances, respectively. The open-circuit impedance parameters, by definition, require that the output be open-circuited to measure z_{11} and z_{21}. However for the measurement of z_{12} and z_{22}, the input must be open-circuited. For the open-circuited output case ($i_2 = 0$),

$$z_{11} = \frac{v_1}{i_1} \qquad (1\text{-}80)$$

and

$$z_{21} = \frac{v_2}{i_1}, \qquad (1\text{-}81)$$

while for the open-circuited input case ($i_1 = 0$),

$$z_{12} = \frac{v_1}{i_2} \qquad (1\text{-}82)$$

and

$$z_{22} = \frac{v_2}{i_2}. \qquad (1\text{-}83)$$

The input and output currents in this case are independent variables. The current gain A_I is given by

$$A_I = \frac{i_2}{i_1} \tag{1-84}$$

From Equation (1-79b) and Figure 1-69,

$$v_2 = i_1 z_{21} + i_2 z_{22} = -i_2 R_L.$$

Hence

$$A_I = \frac{i_2}{i_1} = \frac{z_{21}}{z_{22} + R_L} \tag{1-85}$$

Other electrical properties can be calculated from the impedance parameters.

Short-Circuit Admittance Parameters

The generalized equations for the input and output currents of the black box (Figure 1-69) are

$$i_1 = v_1 y_{11} + v_2 y_{12} \tag{1-86}$$

and

$$i_2 = v_1 y_{21} + v_2 y_{22}, \tag{1-87}$$

where v_1 and v_2 are independent variables and y_{11}, y_{12}, y_{21}, and y_{22} are the input, reverse transfer, forward transfer, and output admittances, respectively. Since these equations describe the short-circuit admittance parameters, it suffices to short-circuit the output ($v_2 = 0$) in order to measure y_{11} and y_{21}. Thus

$$y_{11} = \frac{i_1}{v_1} \tag{1-88}$$

and

$$y_{21} = \frac{i_2}{v_1}. \tag{1-89}$$

To measure the y_{12} and y_{22}, it suffices to short-circuit the input ($v_1 = 0$). Therefore

$$y_{12} = \frac{i_1}{v_2} \tag{1-90}$$

and

$$y_{22} = \frac{i_2}{v_2}. \tag{1-91}$$

All other electrical properties of the black box can be calculated from these admittance parameters.

Hybrid Parameters

Since it is easier to virtually open-circuit a low-impedance circuit than to open-circuit a high-impedance one, the impedance parameters are most useful to describe low-impedance devices and/or circuits. Likewise, since it is easier to virtually short-circuit a high-impedance circuit than a low-impedance one, the admittance parameters are most useful to describe high-impedance devices and/or circuits. It must be realized that when we talk of open-circuit or short-circuit we are speaking only of ac signal frequencies and that the necessity of applying dc biases prevents the application of actual physical short or open circuits.

The hybrid, or *h*, parameters have been found to provide the most useful tool in modern transistor circuit analysis. This is mainly a result of the fact that the *h* parameters are a combination of impedance and admittance parameters ideally fitting the low input and high output impedances of the modern transistor. For the *h* parameters, the black-box equations read

$$v_1 = i_1 h_{11} + v_2 h_{12} \qquad (1\text{-}92a)$$

and

$$i_2 = i_1 h_{21} + v_2 h_{22}, \qquad (1\text{-}92b)$$

where h_{11}, h_{12}, h_{21}, and h_{22} are the input impedance, reverse transfer voltage ratio, forward transfer current ratio, and output admittance, respectively. When the output is short-circuited ($v_2 = 0$),

$$h_{11} = \frac{v_1}{i_1} \qquad (1\text{-}93)$$

and

$$h_{21} = \frac{i_2}{i_1}, \qquad (1\text{-}94)$$

and when the input is open-circuited ($i_1 = 0$),

$$h_{12} = \frac{v_1}{v_2} \qquad (1\text{-}95)$$

and

$$h_{22} = \frac{i_2}{v_2}. \qquad (1\text{-}96)$$

The h_{11}, h_{21}, and h_{22} terms approximate the actual typical operating conditions, even though the latter do not occur with either input or output terminals shorted or open.

In the hybird small-signal transistor model, the h parameters are referred to as

$$h_i = \text{input impedance } (= h_{11})$$
$$h_f = \text{forward current transfer ratio } (= h_{21})$$
$$h_r = \text{reverse voltage transfer ratio } (= h_{12})$$
$$h_0 = \text{output admittance } (= h_{22}).$$

The common-emitter parameters are h_{ie}, h_{fe}, h_{re}, and h_{oe}; the common-base parameters are h_{ib}, h_{fb}, h_{rb}, and h_{ob}; and the common-collector parameters are h_{ic}, h_{fc}, h_{rc}, and h_{oc}.

Figure 1-70 shows the hybrid small-signal model, which is valid in the active region of the transistor, for any configuration at low frequencies. Here V_i is the input voltage, V_L is the output or load voltage, I_i is the input current, I_L is the output current, and Z_L is the impedance loading the output. Notice that this model contains two dependent sources $h_r V_L$ and $h_f I_i$. From Figure 1-70b we have

$$I_L = \frac{-h_f I_i / h_0}{1/h_0 + Z_L} = \frac{-h_f i_i}{1 + h_0 Z_L}.$$

Hence the current gain is

$$A_I \equiv \frac{I_L}{I_i} = \frac{-h_f}{1 + h_0 Z_L}. \tag{1-97}$$

The input impedance is

$$Z_i \equiv \frac{V_i}{I_i} = h_i + h_r A_I Z_L. \tag{1-98}$$

The voltage gain is

$$A_v \equiv \frac{V_L}{V_i} = \frac{A_I Z_L}{Z_i}. \tag{1-99}$$

Equation (1-99) does not contain the h parameters explicitly and hence is valid regardless of what equivalent circuit is used for the transistor.

The negative of the current transfer ratio with the output short-circuited (often referred to simply as the short-circuit current gain) for the *CE* configuration is often called the beta of the transistor or $h_{fe} = \beta$. For the *CB* configuration this quantity is called the alpha of the transistor, or $h_{fb} = -\alpha$.

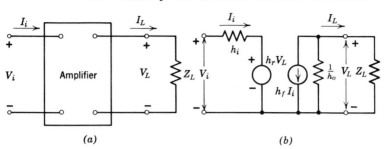

(a)   (b)

Fig. 1-70   (a) A transistor amplifier in either the CE, CB, or CC configuration; (b) the hybrid-equivalent circuit for low-frequency small-signal variations from the quiescent operating point.

Table 1-2 gives the approximate conversion formulas of the h parameters and T-equivalent circuit, in addition to the typical values for the transistor 2N525 at $I_E = 1$ mA.

From Table 1-2 we find that the CC parameters are given in terms of the CE hybrid values by the following nearly exact relationships:

$$h_{ic} = h_{ie}, \qquad h_{fc} = -(h_{fe} + 1),$$
$$h_{rc} = 1 - h_{re} \simeq 1 \qquad \text{(since } h_{re} << 1\text{),}$$
$$h_{oc} = h_{oe}.$$

Therefore, the current gain A_I, input impedance Z_i, and voltage gain A_v for the emitter follower are given by the following equations:

$$A_I = \frac{h_{fe} + 1}{1 + h_{0e} Z_e}, \tag{1-97a}$$

$$Z_i = h_{ie} + A_I Z_e, \tag{1-98a}$$

$$1 - A_v = \frac{h_{ie}}{Z_i}. \tag{1-99a}$$

Here Z_e is the load in the emitter. The voltage gain of an emitter is always less than 1.

1-13   THE HYBRID-π COMMON-EMITTER MODEL – THE HIGH-FREQUENCY SMALL-SIGNAL EQUIVALENT CIRCUIT

One of the most popular high-frequency small-signal equivalent circuits is the hybrid-π common-emitter model shown in Figure 1-71. All parameters (resistances and capacitances) in this model are assumed to be independent of frequency. The base spreading resistance r_{bb}' is the actual base resistance appearing between the active region (point B') and

Table 1-2
Approximate Conversion Formulas of h Parameters

Symbols IRE	Symbols Other	Common Emitter	Common Base	Common Collector	T Equivalent Circuit (Approximate)
h_{ie}	$h_{11e}, \dfrac{1}{Y_{11e}}$	1400 ohms	$\dfrac{h_{ib}}{1+h_{fb}}$	h_{ic}	$r_b + \dfrac{r_e}{1-a}$
h_{re}	$h_{12e}, \mu_{bc'}$ μ_{re}	3.37×10^{-4}	$\dfrac{h_{ib}h_{ob}}{1+h_{fb}} - h_{rb}$	$1 - h_{rc}$	$\dfrac{r_e}{(1-a)r_c}$
h_{fe}	h_{21e}, β	44	$-\dfrac{h_{fb}}{1+h_{fb}}$	$-(1+h_{fc})$	$\dfrac{a}{1-a}$
h_{oe}	$h_{22e}, \dfrac{1}{Z_{22e}}$	27×10^{-6} mhos	$\dfrac{h_{ob}}{1+h_{fb}}$	h_{oc}	$\dfrac{1}{(1-a)r_c}$
h_{ib}	$h_{11}, \dfrac{1}{Y_{11}}$	$\dfrac{h_{ie}}{1+h_{fe}}$	31 ohms	$-\dfrac{h_{ic}}{h_{fc}}$	$r_e + (1-a)r_b$
h_{rb}	$h_{12}, \mu_{ec'}$ μ_{rb}	$\dfrac{h_{ie}h_{oe}}{1+h_{fe}} - h_{re}$	5×10^{-4}	$h_{rc} - 1 - \dfrac{h_{ic}h_{oc}}{h_{fc}}$	$\dfrac{r_b}{r_c}$
h_{fb}	h_{21}, a	$-\dfrac{h_{ie}}{1+h_{fe}}$	-0.978	$-\dfrac{1+h_{fc}}{h_{fc}}$	$-a$
h_{ob}	$h_{22}, \dfrac{1}{Z_{22}}$	$\dfrac{h_{oe}}{1+h_{fe}}$	0.60×10^{-6} mhos	$-\dfrac{h_{oc}}{h_{fc}}$	$\dfrac{1}{r_c}$
h_{ic}	$h_{11c}, \dfrac{1}{Y_{11c}}$	h_{ie}	$\dfrac{h_{ib}}{1+h_{fb}}$	1400 ohms	$r_b + \dfrac{r_a}{1-a}$
h_{rc}	$h_{12c}, \mu_{be'}$ μ_{rc}	$1 - h_{re}$	1	1.00	$1 - \dfrac{r_e}{(1-a)r_c}$
h_{fc}	h_{21c}, a_{eb}	$-(1+h_{ie})$	$-\dfrac{1}{1+h_{fb}}$	-45	$-\dfrac{1}{1-a}$
h_{oc}	$h_{22c}, \dfrac{1}{Z_{22c}}$	h_{oe}	$\dfrac{h_{ob}}{1+h_{fb}}$	27×10^{-6} mhos	$\dfrac{1}{(1-a)r_c}$
	a	$\dfrac{h_{fe}}{1+h_{fe}}$	$-h_{fb}$	$\dfrac{1+h_{fc}}{h_{fc}}$	0.978
	r_c	$\dfrac{1+h_{fe}}{h_{oe}}$	$\dfrac{1-h_{rb}}{h_{ob}}$	$-\dfrac{h_{ic}}{h_{oc}}$	1.67 MΩ
	r_e	$\dfrac{h_{re}}{h_{oe}}$	$h_{ib} - \dfrac{h_{rb}}{h_{ob}}(1+h_{fb})$	$\dfrac{1-h_{rc}}{h_{oc}}$	12.5 ohms
	r_b	$h_{ie} - \dfrac{h_{re}}{h_{oe}}(1+h_{fe})$	$\dfrac{h_{rb}}{h_{ob}}$	$h_{ic} + \dfrac{h_{fc}}{h_{oc}}(1-h_{rc})$	840 ohms

Source: "Transistor Manual," 7th ed., General Electric Company, 1964.
Numerical values are typical for the 2N525 at 1 mA, 5 volts.

the external base contact (point B). The transistor transconductance g_m is linearly related to the emitter current I_E and varies inversely with the absolute temperature T as follows:

$$g_m = \frac{h_{fe}}{1 + h_{fe}} \frac{|I_E|}{nV_T} \quad \text{mhos,} \tag{1-100}$$

where n equals 1 for germanium and approximately 2 for silicon and where $V_T = (T/11,600)$ volts (Sec. 1-8). Thus, for a germanium transistor at room temperature (with $h_{fe} \gg 1$), with I_E in milliamperes,

$$g_m \simeq \frac{|I_E|}{26} \quad \text{mhos.} \tag{1-101}$$

If, in the low-frequency hybrid model, the CE h parameters h_{fe}, h_{ie}, h_{re}, and h_{oe} are determined at a given emitter current I_E, then the resistances in the hybrid-π circuit can be calculated from the following four equations, in the order given:

$$r_{b'e} = \frac{1}{g_{b'e}} = \frac{h_{fe}}{g_m}, \tag{1-102}$$

$$r_{bb'} = h_{ie} - r_{b'e}, \tag{1-103}$$

$$r_{b'c} = \frac{1}{g_{b'c}} = \frac{r_{b'e}}{h_{re}}, \tag{1-104}$$

$$g_{ce} = \frac{1}{r_{ce}} = h_{oe} - (1 + h_{fe}) g_{b'c}. \tag{1-105}$$

The emitter junction capacitance is given by

$$C_e = C_{b'e} \simeq \frac{g_m}{2\pi f_T}, \tag{1-106}$$

where f_T is the frequency at which the CE short-circuit current gain

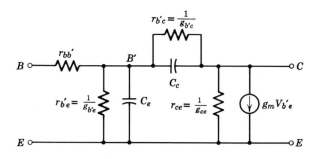

Fig. 1-71   Hybrid-π common emitter model.

drops to unity. The collector junction capacitance $C_c = C_{b'c}$ is the measured CB output capacitance with the input open ($I_E = 0$). Typical values for these capacitances are

$$C_e = 100 \text{ pF} \qquad C_c = 3 \text{ pF}.$$

1-14   OPERATIONAL AMPLIFIERS – ONE IMPORTANT APPLICATION OF DC AMPLIFIERS

Principle

An operational amplifier is a voltage-feedback dc amplifier which makes available a virtual short-circuit input or virtual ground input, as shown in Figure 1-72. The term "virtual" is used to imply that, while the feedback from output to input through Z' serves to keep the voltage V_i at zero, no current actually flows through this short. In Figure 1-72c the virtual ground is represented by the heavy double-headed arrow. The current from the generator V_s continues past this virtual short through the impedance Z', so that $V_0 = -IZ'$. This equation will be derived from the equivalent circuit shown in Figure 1-74.

In Figure 1-72a an amplifier, with input terminals 1 and 2 and output terminals 3 and 4, whose gain is negative, real, and large has been augmented by the addition of two impedances Z and Z'. The impedance Z_i represents the input impedance of the amplifier. The base amplifier within the box may consist of one or more transistor or vacuum-tube stages in cascade.

In order to discuss the equivalent circuit of the operational amplifier, it is convenient to introduce Miller's theorem here. Consider the arbitrary circuit configuration in Figure 1-73. In Figure 1-73a a set of terminals 1-2 is chosen in N_1 and a second set 3-4 is designated in N_2, so that 1 and 3 are connected through Z' and 2 and 4 through a short circuit. The voltage ratio V_2/V_1 is designated by K. We shall now show that the current I_1 drawn from N_1 can be obtained by disconnecting terminal 1 from Z' and simply bridging an impedance $Z'/(1 - K)$ across 1-2, as indicated in Figure 1-73b. The current I_1 is given by

$$I_1 = \frac{V_1 - V_2}{Z'} = \frac{V_1(1-K)}{Z'} = \frac{V_1}{Z'/(1-K)} = \frac{V_1}{Z_1}. \qquad (1\text{-}107)$$

Therefore, if $Z_1 \equiv Z'/(1 - K)$ were shunted across terminals 1-2, the current I_1 drawn from N_1 would be the same as that from the original circuit. Accordingly, as far as N_1 is concerned, Figures 1-73a and b are equivalent. Similarly, the current I_2 drawn from N_2 may be calculated by

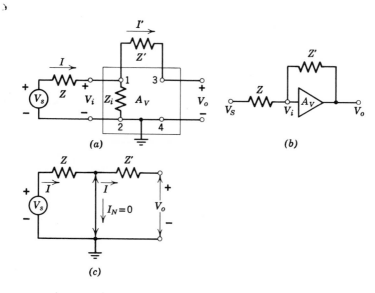

(a)

(b)

(c)

Fig. 1-72 (a) Representation of an operational amplifier; (b) simplified representation of the amplifier; (c) virtual ground in the amplifier.

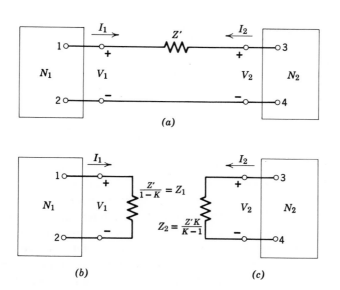

(a)

(b)

(c)

Fig. 1-73 Pertaining to Miller's theorem. $K \equiv V_2/V_1$.

81

replacing the connections to the terminals 3-4 by an impedance Z_2, which is given by

$$Z_2 \equiv \frac{Z'}{1 - 1/K} = \frac{Z'K}{K - 1}. \tag{1-108}$$

Therefore, as far as N_2 is concerned, Figures 1-73a and c are equivalent.

Now let us return to the operational amplifier and define $A_v \equiv V_0/V_i$ to be the voltage amplification with Z' in place. Comparing Figure 1-72a with Figure 1-73, we see that Miller's theorem is directly applicable with $K \equiv A_v$. Therefore, an equivalent circuit of the operational amplifier, which gives the same input current I from the source V_s, the same amplifier input voltage V_i, and consequently the same output voltage V_0 as in Figure 1-72a, is indicated in Figure 1-74.
From Figure 1-74 we see that if

$$\left| \frac{Z'}{1 - A_v} \right| << |Z_i|, \tag{1-109}$$

then $I' \simeq I$. Under these conditions the output voltage is

$$V_0 = A_v V_i = A_v I \frac{Z'}{1 - A_v}. \tag{1-110}$$

As $|A_v| \to \infty$ the impedance across terminals 1 and 2 approaches zero (a short-circuit), and $I \simeq V_s/Z$. Also, as $|A_v| \to \infty$, the output is

$$V_0 \simeq - IZ' = -\frac{Z'}{Z} V_s,$$

and the overall voltage gain is

$$A_f \equiv \frac{V_0}{V_s} = -\frac{Z'}{Z}. \tag{1-111}$$

Therefore, if the amplifier has adequate gain, the overall amplification depends only on the impedances Z' and Z and not on other elements or on the characteristics of the active devices (transistors or vacuum tubes) used in the base amplifier. The operational amplifier suppresses the effect of active device nonlinearity and the effects of the variability of gain with operating point, age, etc. These features of stability, as well as other useful features which are characteristic of the operational amplifier, result from the fact that this amplifier configuration incorporates feedback. The input signal to the amplifier V_i in Figure 1-72a is a linear combination of the external signal V_s and the output signal V_0. Since the signal which is fed back to the input is proportional to the output voltage,

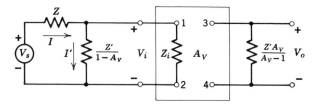

Fig. 1-74   Equivalent circuit of operational amplifier.

the amplifier is described as incorporating voltage feedback.[1] The feedback is negative (or degenerative) in the sense that the gain with feedback $|A_f|$ is less than the gain without feedback $|A_v|$. The "operational" amplifier is so called because it may be used to perform many mathematical operations. Among the basic configurations of operational amplifiers are the operational integrator, operational differentiator, sign changer (or inverter), scale changer, and phase shifter.

Integrating Amplifiers

The operational integrator is shown in Figure 1-75. The double-headed arrow in Figure 1-75*b* represents a virtual ground. The resistor R is used for Z and the capacitor C for Z'. The input is designated by $v = v(t)$ and the current by $i = i(t)$. From the equivalent circuit of Figure 1-75 we see that $i = v/R$ and

$$v_0 = -\frac{1}{C} \int i\, dt = -\frac{1}{RC} \int v\, dt. \qquad (1\text{-}112)$$

Hence the output is proportional to the integral of the input.

The basic requirements for an operational integrator are a large open-loop current amplification and a 180° phase reversal, which can be conveniently achieved by using an odd number of transistor amplifier stages in the common-emitter connection. In general, the number of stages, however, is restricted by the bandwidth and by stability requirements. Therefore, in each case, a compromise should be reached between gain, stability, and bandwidth.

Figure 1-76 shows the circuit diagram of an integrating amplifier. The silicon diodes D_1 and D_2 in this circuit prevent the integrator from overloading. Within a certain output voltage range, these diodes are biased in the reverse direction. If the output voltage does not fall within this range, one of the diodes starts conducting and the integrator becomes a

---

[1] In a current feedback amplifier the feedback voltage is proportional to the output current rather than to the output voltage.

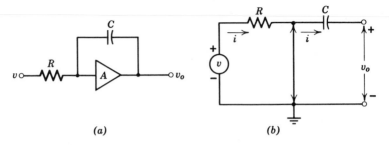

Fig. 1-75   (a) Operational integrator; (b) equivalent circuit.

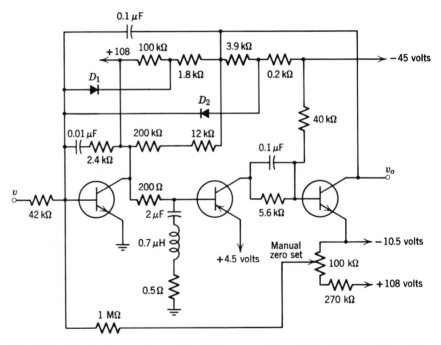

Fig. 1-76   Circuit diagram of operational integrator. (From Richard F. Shea, *Transistor Circuit Engineering*, Wiley, New York, 1957.)

dc amplifier with a voltage attenuation of approximately 10:1. The diodes should have very large reverse resistances and very small shunt capacitances.

Operational Differentiator

If a capacitor C is used for Z and a resistor R for Z', then an operational differentiator is formed. From the equivalent circuit of Figure 1-77 we see that $i = C \, dv/dt$ and

$$v_0 = -Ri = -RC \frac{dv}{dt}. \tag{1-113}$$

Therefore, the output is proportional to the derivative of the input.

Other Basic Configurations

SIGN CHANGER (PHASE INVERTER). If $Z = Z'$, then $A_f = -1$, and the sign of input signal has been changed. Therefore, if two such amplifiers are connected in cascade, the output from the second stage equals the signal input without change of sign.

SCALE CHANGER. If the ratio $Z'/Z = k$, a real constant, then $A_f = -k$ and the scale has been multiplied by a factor $-k$.

PHASE SHIFTER. Suppose Z and Z' are equal in magnitude but differ in angle. Then the operational amplifier shifts the phase of a sine-wave input and at the same time preserves its amplitude.

1-15 SOME OTHER APPLICATIONS OF DC AMPLIFIERS

Difference Amplifiers

Single-ended or unbalanced signals occur on a single terminal with reference to ground. Paired signals (called balanced, push-pull, or double-ended) occur as separate voltages (or currents) v_1 and v_2 on two terminals, with $v_1 = -v_2$. An amplifier which is capable of responding only to the difference $v_1 - v_2$ between two inputs is called a difference amplifier, as shown in Figure 1-78a. Conversely, one which can deliver two equal and opposite outputs from a single input is called a phase splitter, as shown in Figure 1-78b.

Difference amplifiers are needed so that signals other than strictly

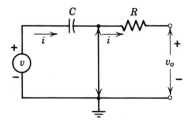

Fig. 1-77 Equivalent Circuit of the operational differentiator.

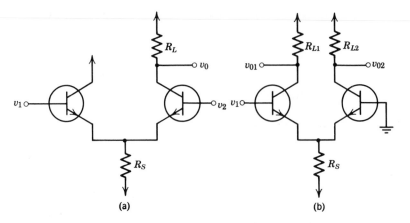

Fig. 1-78   (*a*) Difference amplifier; (*b*) phase splitter.

push-pull signals can be accepted. It is sometimes desirable to subtract interference (for example, induced hum or thermal drifts) from a signal; this can be done in the difference amplifier by means of a dummy signal lead which contains much the same interference but no true signal. In these situations the inputs v_1 and v_2 are not truly balanced. We may resolve them into a difference signal v_d and a common-mode component v_c as follows:

$$v_d = v_1 - v_2, \qquad v_c = \frac{v_1 + v_2}{2}. \tag{1-114}$$

The difference amplifier should respond to v_d and ignore v_c as much as possible. This can be expressed as a requirement for a high rejection ratio, which is defined as the ratio of gains for v_d and v_c. Hence if the rejection ratio is high enough, the output can be given by

$$v_o \simeq K(v_1 - v_2), \tag{1-115}$$

where K is a constant which depends on the gain of the transistors used in the circuit. Because of the common-mode effect,[1] there is usually a small voltage Δv present across R_L when $v_1 = v_2$. However, this voltage ordinarily introduces an error of only about 1 percent or less.

A difference amplifier may be used as a means of stabilizing the output and input impedances of a dc amplifier, as well as for reducing the equivalent input drift. Figure 1-79 shows the basic principle of such an

[1]The common-mode effect may be defined as the ratio of the signal voltage Δv present across R_L when $v_1 = v_2$ to the voltage that appears across R_L when either v_1 or v_2 alone is zero.

arrangement, where the difference amplifier is used as a first stage of the dc amplifier. Figure 1-80 shows a circuit using silicon transistors and employing difference circuits in the first and second stages of the dc amplifier.

The difference amplifier can be viewed as a special type of differential

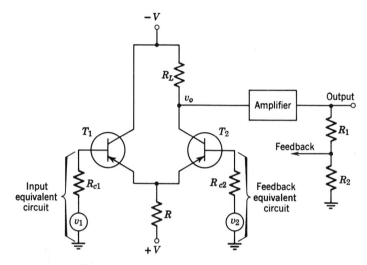

Fig. 1-79   Basic principle of difference amplifier.

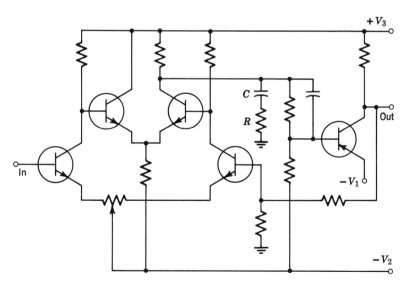

Fig. 1-80   Stabilized dc amplifier employing difference circuits.

amplifier. If, in Figure 1-78a, the collector of the first transistor connects a load resistor which is similar to that in conjunction with the second transistor, the circuit becomes a general type of differential amplifier. The differential output voltage is the difference of the two collector outputs.

Examples of Stabilized DC Amplifiers Used in the Typical Power Supplies

Figure 1-81 shows the stabilized dc amplifiers used in the typical power supplies. The + and −7-, the + and −10- and the + and −21-V power supplies are all of the same type: a series regulator (Q_1, Q_3, Q_5, Q_7, Q_9, or Q_{11}) driven by an emitter follower (Q_2, Q_4, Q_6, Q_8, Q_{10}, or Q_{12}), with a 5-percent Zener diode as a reference element. This reference element is placed in the base circuit of the emitter follower so that the error or difference signal between the reference voltage and some portion of the output voltage is developed. The regulated output will approximate the reference voltage.

The +48-V supply is switched on during PRINT with $CR35$ and during PUNCH or READ-IN with $CR36$. The base of Q_{16} will be clamped to +7 V with $CR37$, and as Q_{16} and Q_{17} are emitter-coupled, the base of Q_{17} will try to be at the same potential, which brings the output to +48 V. Note that this +48-V must be loaded with at least 15 mA to prevent oscillations. The collector of Q_{16} is unloaded with a complementary emitter follower (Q_{14} and Q_{15}) to drive the base of the series regulator Q_{13}. In the complementary emitter follower, one of the transistor pair is an n-p-n unit and the other is a p-n-p unit, and these two units have similar current and voltage ratings. An emitter-follower circuit using such a pair serves to provide powerful drive in both directions.

The −75 V is stabilized with a Zener diode ($CR45$) and a differential amplifier, Q_{22} and Q_{23}. The trim potentiometer $R27$ is used to set the output voltage at −75 V.

The collector of Q_{23} drives a series regulator Q_{18} through a complementary emitter follower Q_{19} and Q_{20}.

Q_{22} and $CR44$, $CR43$ protect the supply against overloading when the voltage across $R24$ exceeds the voltage across $CR44$ and $CR43$. Q_{21} starts turning "on," and will therefore block Q_{18}.

The characteristics of the Zener diodes and transistors used in the circuit of Figure 1-81 are given in Tables 1-3 and 1-4, respectively.

1-16 THE BASIC TRANSISTOR SWITCHING CIRCUIT

Switching Principle

Transistors have become the most useful, versatile, and widely used devices in switching applications. The basic concept in any switching

Table 1-3
Characteristics of Silicon Zener Diodes

Type Number	Reference Voltage Range Minimum E_{b_1}, (V)	Maximum E_{b_2}, (V)	Nominal Toler. $\pm 0/0$	@I_z, (mA)	Dynamic Impedance Z, (Ω)	@I_z, (mA)	Maximum Dissipation (mW)	Nominal Temperature Coefficient, (%/°C)	Maximum Temperature, (°C)
1N958	6.0	9.0	5.0	16.5	5.5	16.5	400	0.045	175J
1N959B	7.79	8.61	5.0	15.0	700	0.50	400	0.048	175A
1N962B	10.45	11.55	5.0	11.5	700	0.25	400	0.06	175A
1N969	17.6	26.4	5.0	5.6	29	5.6	400	0.08	175J

Table 1-4
Characteristics of Transistors

Type	Use	Maximum Ratings P_c @ 25°C (mW)	BV_{CE} $BV_{CB}*$, (V)	BV_{EB}, (V)	I_c, (mA)	T_J, °C	Electrical Parameters Minimum @I_c, $h_{fe}-h_{FE}*$	(mA)	Minimum f_{hfb}, (MHz)	Maximum I_{CO}, (µA)	@V_{CB}
Ge Si*											
2N398, PNP	on-off HV,LW	50	−105	−50	−110	85	20*	−5 mA	−	−14	−2.5
2N404, PNP	sw	120	− 24	−12	−100	85	−	−	4	− 5	−12
2N914, NPN*	sw	360	40	5		200J	30T	−	400	.025	−
2N1038, PNP		20W	− 40		−3A	95	35*	−1A	−	−125	.5
2N1304, NPN	sw	300	20		300	100	40*	10	5	6	25
2N1540, PNP	power	150W	− 60*		−5A		50*	−3A	−	−	−

1. BV_{CE}, BV_{CEO}: dc-breakdown voltage, collector to emitter, with base open-circuited. Specify I_C.

2. BV_{CB}, BV_{CBO}: dc breakdown voltage collector to base junction reverse biased, emitter open-circuited. Specify I_C.

3. BV_{EB}, BV_{EBO}: dc breakdown voltage, emitter to base, with collector open-circuited.

4. f_{hfb}: (common base) small-signal short-circuit forward-current transfer-ratio cutoff frequency.

5. h_{fe}: (common emitter) small-signal short-circuit forward-current transfer ratio, output ac short-circuited.

6. h_{FE}: (common emitter) static value of forward current transfer ratio, $h_{FE} = I_C/I_B$.

7. I_{CO}, I_{CBO}: dc collector current when collector junction is reverse biased and emitter is open-circuited.

8. T_J: junction temperature.

9. J: operating junction temperature.

10. T: typical values.

circuit is a discrete change of state, usually a voltage change or a current change or both. This change of state may be used to perform logical functions, as in a computer, or to transfer energy, as in relay drivers. Transistor switching circuits act as generators, amplifiers, inverters, frequency dividers, and wave shapers to provide limiting, triggering, gating, and signal-routing functions. These applications are normally

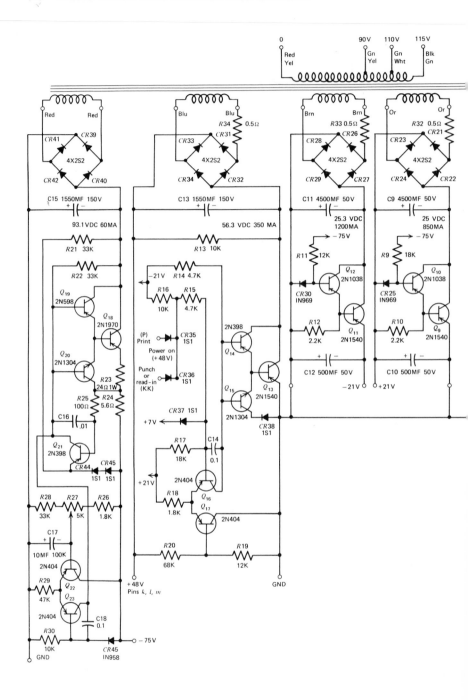

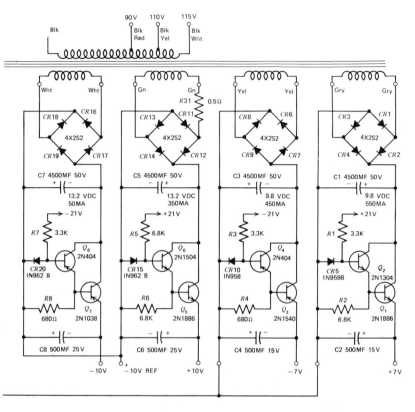

Fig. 1-81 Examples of stabilized dc amplifiers used in typical power supplies. (From *Operating and Maintenance Instruction for 400 Channel Analyzer System Model* 2063 1/2, Radiation Counter Laboratories, Inc.)

characterized by large-signal or nonlinear operations of the transistor.

Transistors in switching circuits are used as switches that are either on (conducting) or off (nonconducting). Depending on the load and voltage, the operating characteristic may be of the three types: saturated mode, current mode, and avalanche mode, as in Figure 1-82. In all three types, the collector current is made to swing from zero (cutoff, maximum collector voltage) to some "on" value. But in the saturated mode, the collector voltage swings almost to emitter voltage; in the current mode, the collector voltage does not swing so far. In the avalanche region, where the collector current rises fast and becomes large due to avalanche effect, a large current can be obtained without dropping the collector

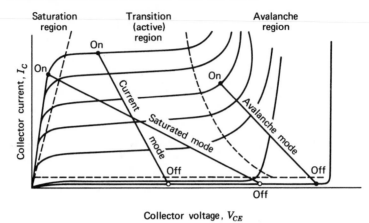

Fig. 1-82   Operating regions for switching modes. (From Motorola *Switching Transistor Handbook*.)

voltage very much. The current-mode operation has the advantage of being capable of faster operation than the more commonly used saturated mode. Regions of operation are similar for all transistor configurations used as switches. Table 1-5 gives the collector-to-emitter saturation voltage [V_{CE} (sat)], the base-to-emitter saturation voltage [V_{BE} (sat) ≡ V_σ], the base-to-emitter voltage in the active region [V_{BE} (active)], at cut-in [V_{BE} (cut-in) ≡ V_γ], and at cutoff [V_{BE} (cutoff)]. The entries in the table are appropriate for an *n-p-n* transistor. For a *p-n-p* transistor the signs of all entries should be reversed.

Table 1-5
Typical *n-p-n* Transistor-Junction Voltages (V) at 25°C

V_{CE} (sat)	V_{BE} (sat) ≡ V_σ	V_{BE} (active)	V_{BE} (cut-in) ≡ V_γ	V_{BE} (cutoff)
Si 0.3	0.7	0.6	0.5	0.0
Ge 0.1	0.3	0.2	0.1	−0.1

When a transistor switching circuit is ON, the resistance should be as low as possible across the transistor to avoid loss of power across the switch. To achieve this low resistance, it is necessary that the transistor be in the saturation region. Enough base current must be supplied to assure that saturation is maintained under "worst-case" operating conditions. In the OFF operation, the impedance across the transistor should be as high as possible. In large-signal operation, the transistor acts as an over-driven amplifier which is driven from the cutoff region to the saturation region. Referring to the simple transistor-switching circuit shown in Figure 1-83, we can explain the large-signal operation as follows: when the switch S_1 is in the OFF position, the emitter-base junction of the transistor is reverse-biased by battery V_{B2} through the current-limiting resistor R_2. The transistor is then in the OFF (cutoff) state. Since both junctions are reverse-biased, the transistor is in normal quiescent conditions. When S_1 is in the ON position, forward bias is applied to the emitter-base junction by battery V_{B1} through the current-limiting resistor R_1. The base current and collector current then increase rapidly until the transistor reaches saturation. The transistor is saturated when the collector current reaches a value at which it is limited by R_3 and V_{CC}. Collector current is then approximately equal to V_{CC}/R_3, and further increases in base drive produce no further increase in collector current. In the saturation region, the collector current is usually at a maximum and collector voltage at a minimum. This value of collector voltage is called the collector-to-emitter saturation voltage $[V_s = V_{CE} \text{ (sat)}]$ and is about 0.1 to 1.0 V, depending on the base current. A transistor operating in the saturation region is in the ON (conducting) state. Both junctions are then forward-biased.

The Transistor as a Switch

THE TRANSISTOR SWITCH.   In the switching circuit of Figure 1-84a the base is connected to the supply voltage through R, so that the transistor

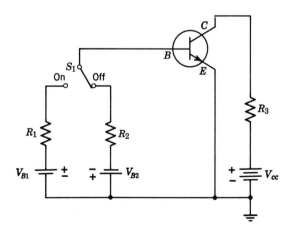

Fig. 1-83 Simple Switching Circuit.

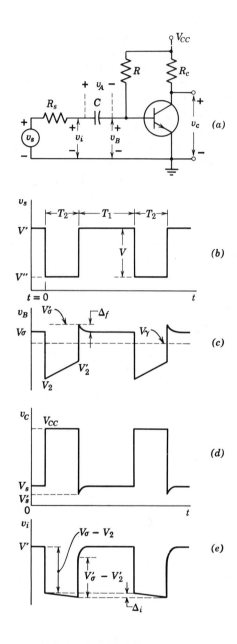

Fig. 1-84 (a) A basic transistor switching circuit; (b) input waveform; (c) base waveform; (d) collector waveform; (e) waveform v_i at junction of R_s and C.

is saturated in the absence of a signal. The square wave v_s in Figure 1-84b is applied to the base through the capacitor C from a source of resistance R_s. The transistor is viewed as a "gate" which opens and closes, and the input waveform is the agency through which the transistor is operated. Hence the input signal is here called a gating waveform, or a train of negative gating pulses, rather than a square wave.

BASE WAVEFORM.   The base circuit of Figure 1-84a is viewed as a diode clamping circuit. The waveform at the base will appear as in Figure 1-84c, which should be compared with Figure 1-53c. In these two figures the time constant with which the capacitor C charges at the positive extremity of the signal is small relative to the interval T_1, and the base voltage returns to the level at which it would remain constantly in the absence of an input signal. This saturation (clamping) level $V_\sigma \equiv V_{BE}$ (sat) will ordinarily be approximately 0.3 or 0.7 V for a germanium or silicon transistor (see Table 1-5). We may find the clamping voltage V_σ more precisely using the load-line construction of Figure 1-85.

Consider the voltage levels of the base waveform v_B of Figure 1-84c. Since at $t = 0-$ the signal v_s is at the level V' and the base is at the saturation level V_σ, then the capacitor voltage is $V' - V_\sigma$. At $t = 0+$, $v_s = V''$, and the capacitor voltage remains $V' - V_\sigma$. Suppose the abrupt change V in v_s is large enough to drive the transistor below the cut-in level V_γ. Then if the small base current at cutoff is neglected, the equivalent circuit from which to calculate v_B can be indicated as in Figure 1-86, where R_{s2} represents the value of the source resistance during the interval T_2, and $v_A = V' - V_\sigma$ immediately after $t = 0$. By using the principle of superposition and remembering that $V = V' - V''$, we can show that the base voltage at $t = 0+$ is given by

$$v_B (0+) = (V_\sigma - V) \frac{R}{R + R_{s2}} + V_{CC} \frac{R_{s2}}{R + R_{s2}} \equiv V_2. \tag{1-116}$$

Notice that (1-116) is valid only if the base is driven below the cut-in level, that is, if $V_2 < V_\gamma$.

Consider the case in which $R_{s2} = R$, so that

$$V_2 = \frac{V_\sigma - V + V_{CC}}{2}. \tag{1-117}$$

The condition $V_2 < V_\gamma$ requires that

$$V > V_{CC} + V_\sigma - 2V_\gamma. \tag{1-118}$$

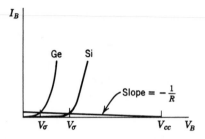

Fig. 1-85   Load-line construction to find clamping voltage at base of Ge or Si transistor.

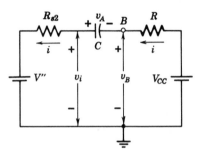

Fig. 1-86   Input circuit equivalent to the base circuit of Figure 1-84a during the time that the transistor is cut off. At $t = 0 +$, $v_A = V' - V_\sigma$.

Since $V_\sigma - 2V_\gamma$ is usually small relative to the supply voltage V_{CC}, we have $V > V_{CC}$, approximately.

During the interval T_2 the capacitor C charges and the voltage at the base rises asymptotically toward V_{CC}. Using (1-2) we can calculate V_2' from

$$V_2' = V_{CC} - (V_{CC} - V_2)\, \epsilon^{-T_2/\tau}, \qquad (1\text{-}119)$$

where $\tau = C(R + R_{s2})$ and $V_2' < V_\gamma$.

When at $t = T_2$ the input rises abruptly by amount V, it carries the base above its initial level V_σ, and an overshoot appears in the v_B waveform, as in Figure 1-84c. To calculate the amplitude Δ_f of this overshoot we first find the change Δv_A in the voltage across the capacitor C in the interval T_2. During the interval T_2 the base side of the capacitor C in Figure 1-84 rises in voltage by the amount $V_2' - V_2$. The current i through R or R_{s2} falls by the amount $(V_2' - V_2)/R$. Therefore the decrease in voltage drop across R_{s2} is $R_{s2}(V_2' - V_2)/R$, and the voltage v_i at the generator side of the capacitor falls by this same amount. The total change in voltage across C is

$$-\Delta v_A = (V_2' - V_2) + \frac{(V_2' - V_2) R_{s2}}{R}$$

or

$$-\Delta v_A = (V_2' - V_2) \frac{1 + R_{s2}}{R}. \tag{1-120}$$

At $t = T_2+$ the circuit is different from its initial state. The difference lies in the fact that the voltage Δv_A exists across the capacitor C and that the source resistance is now R_{s1}. The change from the initial state which results from this voltage change Δv_A can be calculated from the circuit of Figure 1-87. Here the incremental input impedance of the transistor is the base-spreading resistance r_{bb}'. The change in voltage across r_{bb}' is the overshoot. Using (1-120) we find, subject to the condition $R >> r_{bb}'$, that

$$\Delta_f \approx - \frac{\Delta v_A \, r_{bb}'}{R_{s1} + r_{bb}'}$$

or

$$\Delta_f \approx (V_2' - V_2) \frac{R + R_{s2}}{R} \frac{r_{bb}'}{R_{s1} + r_{bb}'}. \tag{1-121}$$

Usually, $R >> R_{s2}$ and $R_{s1} >> r_{bb}'$. In this case

$$\Delta_f \approx (V_2' - V_2) \frac{r_{bb}'}{R_{s1}}. \tag{1-122}$$

The peak value of v_B is $V_\sigma' = V_\sigma + \Delta_f$. The overshoot decays with a time constant

$$\tau' = (R_{s1} + r_{bb}') \, C. \tag{1-123}$$

COLLECTOR WAVEFORM. The collector waveform is shown in Figure 1-84d. During the interval T_2, when the transistor is below cutoff, its collector voltage is $v_C \approx V_{CC}$. At $t = 0-$, when the transistor is in saturation, $v_C \equiv V_s = V_{CE}$ (sat). This saturation voltage is obtained by drawing a load line corresponding to V_{CC} and R_c on the collector characteristics

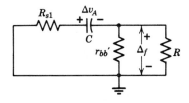

Fig. 1-87   The input equivalent circuit from which to calculate the overshoot Δ_f at $t = T_2+$.

(see Figure 1-66) and locating the intersection for a base current $I_B = (V_{CC} - V_\sigma)/R$. The load line is obtained simply as follows. Assume the collector current is zero; then the collector voltage is the supply voltage V_{CC}. This fixes the point x_m on the X axis. The maximum current that can flow is simply the maximum current that V_{CC} can produce in the collector resistor R_c. This is V_{CC}/R_c; at this point the collector voltage is zero. This fixes the point y_m on the Y axis. Points x_m and y_m determine the load line. A good approximation for V_{CE} (sat) is 0.1 V for a germanium and 0.3 V for a silicon transistor (see Table 1-5). At $t = T_2+$, when the base overshoots, there will be a small undershoot in collector voltage. Because v_C cannot go negative, V_s' differs from V_s by a maximum of 0.1 V for Ge and 0.3 V for Si.

WAVEFORM v_i AT JUNCTION OF R_s AND C.   At $t = 0-$ there is no current in C, and so the upper level of v_i is identical with the upper level V' of v_s. At $t = 0+$ the voltage v_B drops abruptly by $V_\sigma - V_2$, as does v_i on the other side of C. As C charges, the capacitor current decreases, the voltage drop across R_{s2} becomes smaller, and hence during the interval T_2 the voltage v_i falls. From (1-120), the change Δ_i in v_i is

$$\Delta_i = (V_2' - V_2)\frac{R_{s2}}{R}. \tag{1-124}$$

At $t = T_2+$ there is an abrupt jump in v_i by $V_\sigma' - V_2'$ that matches the abrupt jump in v_B on the other side of C. For $t > T_2$ the capacitor discharges through R_{s1} and the input diode. As the capacitor discharges, the voltage v_i rises to its starting point asymptotically with the time constant τ' [see (1-123)] and the collector undershoot, as well as the base overshoot, decays with this same time constant.

Example 1-4

The circuit of Figure 1-84a has $V_{CC} = 10$ V, $R_c = 500\,\Omega$, $R = 40$ K, $C = 0.1\ \mu$F, $R_{s1} = R_{s2} = 10$ K, $V = 10$ V, $T_2 = 1.0$ msec; it uses an n-p-n germanium transistor with characteristics similar to those of the p-n-p type 2N404 in Figure 1-88. Find the voltage levels of all the waveforms of Figure 1-84, assuming $r_{bb}' = 100\,\Omega$.

Solution   Assume the transistor is in saturation; then the base-to-emitter saturation voltage is $V_\sigma = 0.3$ V (see Table 1-5) and the base current is $I_B = (10 - 0.3)/40 = 0.24$ mA. This base current is adequate to keep the transistor in saturation (see Figure 1-88). From (1-116)

$$V_2 = (0.3 - 10)\frac{40}{40 + 10} + 10\frac{10}{40 + 10} = -5.8\ \text{V}.$$

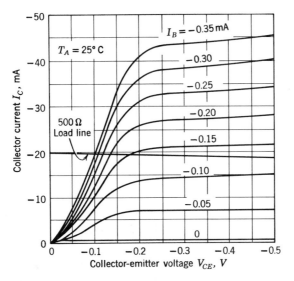

Fig. 1-88   Saturation-region common-emitter characteristics of the 2N404 alloy-junction germanium transistor. A load line corresponding to $V_{CC} = 10$ volts, $R_L = 500$ ohms, is superimposed. (Courtesy of Texas Instruments, Inc.)

From (1-119)

$$V_2{}' = 10 - (10 + 5.8)\, \epsilon^{-1.0/5.0} = -2.9 \text{ V}.$$

From (1-121)

$$\Delta_f = (-2.9 + 5.8) \frac{40 + 10}{40} \frac{0.1}{10.1} = 0.04 \text{ V}.$$

The sum of V_σ and Δ_f is

$$V_\sigma{}' = V_\sigma + \Delta_f = 0.3 + 0.04 \approx 0.3 \text{ V}.$$

   In the waveform v_C the upper level is 10 V. The lower level is V_{CE} (sat) = 0.1 V (see Table 1-5).

   The changes in level in v_i are

$$V_\sigma - V_2 = 0.3 + 5.8 = 6.1 \text{ V}$$

$$V_\sigma{}' - V_2{}' = 0.3 + 2.9 = 3.2 \text{ V}.$$

From (1-124)

$$\Delta_i = (-2.9 + 5.8) \frac{10}{40} = 0.7 \text{ V}.$$

The recovery time constant is

$$\tau' = (R_{s1} + r_{bb}') \, C = (10.1 \times 10^3)(0.1 \times 10^{-6}) \text{ sec} \simeq 1 \text{ msec} = T_2.$$

In order for the transient for $t > T_2$ to die down in the interval T_1, as in Figure 1-84, we must have $T_1 >> \tau'$, or $T_1 >> T_2$.

1-17   BASIC LOGIC CIRCUITS

Introduction to Logic Systems

Logic circuits are logic gates used to implement Boolean equations which are part of an algebraic system for the mathematical analysis of logic. The four basic logic circuits are the OR, AND, NOT, and FLIP-FLOP circuits, which are most commonly employed in a large-scale digital system, such as a computer, or in a data-processing system. A computer is a system which processes and stores very large amounts of data and which solves scientific problems involving highly complex numerical computation when solution by human calculators is not feasible. Figure 1-89 shows the basic elements of a computer: memory, central processor, input, and output. The numbers and the instructions which form the program the computer is to follow are stored in the memory. The central processor consists of the control and the arithmetic units. The control converts the order into an appropriate set of voltages to operate switches, etc., so that the instructions conveyed by the order can be carried out. The arithmetic unit contains the circuits which actually perform the arithmetic computations: addition, subtraction, etc. The input-output devices are used for inserting numbers and orders into the memory and for reading the final result.

A digital system commonly employs transistors which operate at cutoff

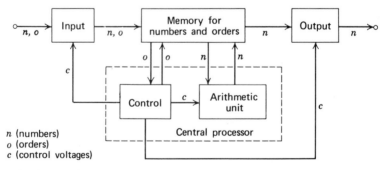

Fig. 1-89   Interrelationship of the various components of an electronic digital computer.

or in saturation, so that the system functions in a binary manner. The binary states are designated by various notations, such as true or false, 1 or 0, etc. The notation true or false often means that the most positive voltage level is the logical "true" state, while the most negative voltage level is the logical "false" state. Binary arithmetic and mathematical manipulation of switching or logic functions are best carried out with two symbols, 0 (zero) and 1 (one). In the binary system of representing numbers[1] the base is 2; only the two numerals 0 and 1 are required to represent a number, and the individual digits represent the coefficients of powers of two. For example, the decimal number 21 is written in the binary representation as 10101, since

$$10101 \equiv 1 \times 2^4 + 0 \times 2^3 + 1 \times 2^2 + 0 \times 2^1 + 1 \times 2^0$$
$$= 16 + 0 + 4 + 0 + 1 = 21.$$

Similarly, decimal numbers such as 0, 1, 2, 5, 10 are written in the binary system as 00000, 00001, 00010, 00101, and 01010, respectively.

The following is the procedure for converting from a decimal to a binary number. Place the decimal number (say 21) on the extreme right of Table 1-6. Next divide by 2 and place the quotient (10) to the left and indicate the remainder (1) directly below it. Repeat this process (for the next column $10 \div 2 = 5$ with a remainder of 0) until a quotient of 0 is obtained. The array of 1's and 0's in the second row is the binary representation of the original decimal number. In this example, decimal 21 = 10101 binary.

Table 1-6
Example of Decimal-to-Binary Conversion

Divide by 2	0	1	2	5	10	21 decimal
Remainder	1	0	1	0	1	Binary

A binary digit (a 1 or a 0) is known as a bit. A group of bits having a significance is a bite, word, or code. In a dc or level-logic system a bit is implemented as one of two voltage levels. If the more positive voltage is the 1 level and the other is the 0 level, the system is said to employ dc positive logic. On the other hand, a dc negative-logic system is one which designates the more negative voltage state of the bit as the 1 level and the more positive as the 0 level. Note that the absolute values of the two voltages are of no significance in these definitions, and the 0 state need

---

[1]The binary system will be easily understood if we refer to the familiar decimal system; for example, the number 1949 (one thousand nine hundred forty nine) has the meaning $1949 \equiv 1 \times 10^3 + 9 \times 10^2 + 4 \times 10^1 + 9 \times 10^0$. Thus the individual digits in a number represent the coefficients in an expansion of the number in powers of 10.

not represent a zero voltage level (although in some systems it might). In a dynamic or pulse logic system a bit is recognized by the presence or absence of a pulse. In a dynamic positive- (negative-) logic system a 1 signifies the existence of a positive (negative) pulse, and a 0 at a particular input or output at a given instant of time indicates that no pulse is present at that particular moment. In a "double-rail" system the variable appears on two leads. A pulse on one lead indicates that the variable has the 0 value, whereas a pulse on the other lead signifies a 1.

Most computers using pulses operate as synchronous systems. To achieve synchronism, there is available in a computer a continuous sequence of pulses of good waveshape whose frequency is usually established by a crystal oscillator known as the master clock. These clock pulses are distributed to all parts of the computer, where they are used to maintain the timing of the system. In a synchronous dynamic system a 1 is implemented by a pulse occurring at the same time as a clock pulse, whereas a signal pulse is absent at a particular clock-pulse time for a 0. Thus a binary number is represented in serial form by a train of pulses, called a binary-coded pulse train, as in Figure 1-90a. Notice that since time increases from left to right, the least-significant pulse occurs at the extreme left (at $t = 0$), whereas in representing a binary number the least-significant bit is placed at the extreme right.

There are two common modes of operation for computers, the serial and the parallel. In the serial mode the pulses (or absence of pulses) occur serially, one after another, and the information (number or instruction) conveyed by this pulse sequence can be transmitted from one place to another over a single communication link (in the simplest case a pair of wires). In the parallel mode each of the pulses (or absence of pulses) needed to represent the information occurs simultaneously on a separate

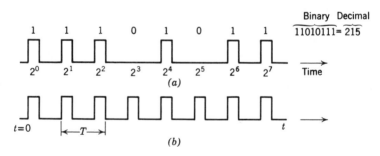

Fig. 1-90 (*a*) A binary-coded pulse train representing a number (or an order) in serial form in a synchronous positive-logic digital system; (*b*) the clock pulses used to maintain the timing of the system.

channel (in the simplest case on a separate wire, there being, say, a common ground), and the information is transmitted in one pulse interval. However as many channels as there are bits in the character may be necessary.

THE OR GATE

DEFINITION AND EXPRESSION. An OR gate has two or more inputs and a single output. It performs the OR operation in accordance with the following definition: *the output of an OR assumes the 1 state if one or more inputs assume the 1 state.* The n inputs to a logic circuit will be designated by $A, B, C, \ldots, N$ and the output by Y. Each of these symbols may assume either 0 or 1. The Boolean expression for an OR gate is

$$Y = A \text{ or } B \text{ or } C \text{ or } \cdots \text{ or } N$$

or

$$Y = A + B + C + \cdots + N. \tag{1-125}$$

where $a +$ between symbols indicates the OR operation.

A DIODE-LOGIC OR CIRCUIT. A diode OR circuit for negative logic is shown in Figure 1-91. The generator source resistance is designated by R_s. Suppose the supply voltage V_R has a value equal to the voltage $V(0)$ of the 0 state for dc logic; then if all inputs are in the 0 state, the voltage across each diode is $V(0) - V(0) = 0$, and none of the diodes conducts. Therefore the output voltage is $v_0 = V(0)$, and Y is in the 0 state. Now if input A is changed to the 1 state, which for negative logic is at the potential $V(1)$, and is less positive than the 0 state, then the diode D_1 will conduct. The output becomes

$$v_0 = V(0) - [V(0) - V(1) - V_\gamma] \frac{R}{R_s + R_f + R},$$

where V_γ is the cut-in voltage,[1] and R_f is the diode forward resistance. Usually $R >> (R_s + R_f)$, and hence $v_0 \simeq V(1) + V_\gamma$. This indicates that the output exceeds the more negative level $V(1)$ by V_γ (approximately 0.2 V for Ge or 0.6 V for Si).

Now, assume that R is much larger than R_s and that both R_f and V_γ are zero. Then, for input A excited, the output is $v_0 = V(1)$ and the gate has performed the following logic: if $A = 1, B = 0, \ldots, N = 0$, then $Y = 1$. For input A excited, the presence of signal sources at $B, C, \ldots, N$

---

[1] In order for a diode to conduct it must be forward-biased by at least the cut-in or breakpoint voltage V_γ (see Figure 1-43 and Section 1-9).

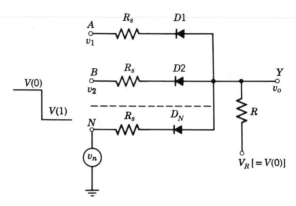

Figure 1-91   A diode OR circuit for negative logic. If all diodes are reversed, this circuit will become a positive-logic OR gate with the output equal to the most positive level $V(1)$ (assuming $R >> R_s$, $R_f = 0$, $V_\gamma = 0$).

does not result in an additional load on generator A. If two or more inputs are in the 1 state, then the output is $V(1)$ and again the OR function is satisfied. If for any reason the level $V(1)$ is not identical for all inputs then the most negative value of $V(1)$ appears at the output, and all diodes except one are nonconducting.

A DYNAMIC LOGIC SYSTEM.   In a dynamic logic system, the output-pulse magnitude is approximately equal to the largest input pulse (regardless of whether the system uses positive or negative logic), and the output pulse will be influenced by the diode capacitance and shunt capacitance. Now, let us see this capacitive effect. Suppose that the level $V(0)$ is at ground potential and that only one generator is supplying an input pulse. Then all diodes but one are reverse-biased during the input pulse, and the capacitance shunted across the output is $C = (n - 1) C_d + C_0$, where n is the number of diodes, C_d is the single diode capacitance, and C_0 is the capacitance across R. The equivalent circuit is shown in Figure 1-92a, where R_s ($<< R$) is the output impedance of the generator furnishing the pulse and the impedance of the other generators is neglected. The input pulse appears at the output with a rounded leading edge whose time constant is $R_s C$. The output waveform is shown in Figure 1-92b. When the input voltage rises at the end of the pulse, the output capacitor will maintain the output voltage and every diode will be reverse-biased, so that the output capacitance will be increased to $C' = nC_d + C_0$. Then C' must discharge through R. Since $RC' >> R_s C$, the trailing-edge rise time will be very much longer than the leading-edge rise time.

When the pulses are at an inconvenient average-voltage level, we can use capacitive coupling in an OR gate, as shown in Figure 1-93. Here the shunt diodes are used for dc restoration. The resistors R' are large and may possibly be omitted altogether with semiconductor diodes. At the termination of the input pulse the corresponding capacitor will quickly discharge through the shunt diode.

AN EMITTER-COUPLED OR GATE. Figure 1-94 shows an OR gate for negative logic using transistors in an emitter-follower configuration. The bottom of the common emitter resistor R goes to a supply voltage equal to $V(0)$. If all inputs are at the 0 level $V(0)$, then each base-to-emitter voltage is at $0V$ $[= V(0) - V(0)]$ and each transistor is virtually at cutoff. Thus the output is $v_0 = V(0)$, or $Y = 0$. On the other hand, if any input is at the 1 level $V(1)$, then, because of emitter-follower action, the output is $v_0 = V(1)$ or $Y = 1$ (neglecting the small base-to-emitter

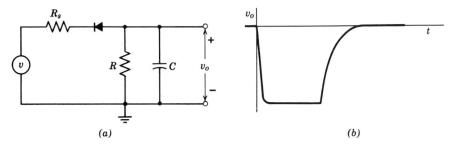

(a)   (b)

Fig. 1-92 (*a*) Equivalent circuit for a diode OR gate with one input excited ($R \gg R_s$); (*b*) output waveform for a negative pulse input.

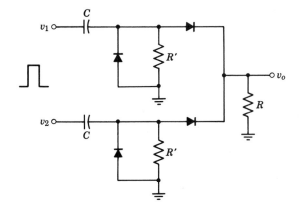

Fig. 1-93 Capacitive coupling employed in a two-input OR gate for positive pulses.

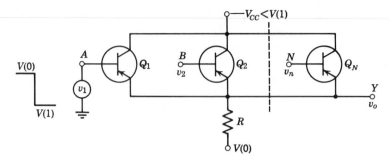

Fig. 1-94   An emitter-coupled OR gate for negative logic. Note that for negative logic the transistors used in this OR gate must be of the *p-n-p* type. (This same circuit is also a positive AND gate.)

voltage). The base-to-emitter voltage of all transistors except the one which is excited is $V(0) - V(1)$. Because of this positive voltage, these transistors are off. Therefore the circuit of Figure 1-94 obeys the OR logic, and has a higher input impedance than the diode-logic-gate. If the excited transistor is to be driven into its active region, the collector supply voltage $-V_{CC}$ must be more negative than $V(1)$. For positive logic, *n-p-n* transistors must be used and V_{CC} must be more positive than $V(1)$. The general appearance of an output pulse from the circuit of Figure 1-94 will be as shown in Figure 1-92. In the active region the output resistance of an emitter follower is approximately $(R_s + h_{ie})/(1 + h_{fe})$, where R_s is the source impedance.

The AND Gate or COINCIDENCE Circuit

DEFINITION AND EXPRESSION.   An AND gate has two or more inputs and a single output, and performs the AND operation in accordance with the following definition: *the output of an AND assumes the 1 state if and only if all the inputs assume the 1 state.* The Boolean expression for an AND gate is

$$Y = A \text{ and } B \text{ and } C \text{ and } \cdots \text{ and } N$$

or

$$Y = A \cdot B \cdot C \cdot \cdots \cdot N. \tag{1-126}$$

where a dot ($\cdot$) between symbols indicates the AND operation.

A DIODE-LOGIC AND CIRCUIT.   The AND gate is also called a *coincidence* circuit. A diode AND gate for negative logic is shown in Figure 1-95*a*. Assume for simplicity that all source resistances R_s are zero and that the diodes are ideal. If any input is at the 0 level $V(0)$, then the diode connected to this input conducts and the output is clamped at the voltage $V(0)$ or $Y = 0$. If all inputs are at the 1 level $V(1)$, then all diodes are reverse-biased and $v_0 = V(1)$ or $Y = 1$. Thus the AND operation has been implemented.

Consider the diode AND gate for positive logic shown in Figure 1-95*b*. If V_R is more positive than $V(1)$, then all diodes will conduct upon a coincidence (all inputs in the 1 state) and the output voltage will be clamped to $V(1)$. In this case the output impedance is low [equal to $(R_s + R_f)/n$ in parallel with R]. If $V_R = V(1)$, then all diodes are cut off at a coincidence, and the output impedance is high (equal to R). If not all inputs have the same upper level $V(1)$, then the output will equal $V(1)_{min}$ [the minimum value of $V(1)$]. If V_R is smaller than all inputs $V(1)$, then all diodes will be cut off upon a coincidence and the output will rise to V_R. If $V_R = V(1)$ and if m inputs are at $V(1)$, then m diodes are reverse-biased and the remaining n-m diodes conduct. Therefore the effective circuit of these diodes in parallel consists of a resistance $(R_s + R_f)/(n - m)$ in series with a voltage V_γ, and the output voltage is given by

$$v_0 = V(1) - [V(1) - V(0) - V_\gamma] \frac{R}{R + (R_s + R_f)/(n - m)}.$$

$$(1\text{-}127)$$

From $(1\text{-}127)^1$ we see that the output will respond to the number m of

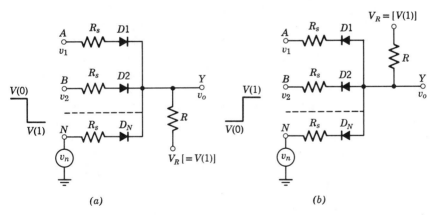

(a)   *(b)*

Fig. 1-95   A diode-logic AND gate for *(a)* negative logic and *(b)* positive logic. The positive AND gate *(b)* is the same circuit as the negative OR gate of Figure 1-91.

¹If all inputs are excited, or $m = n$, then $v_0 = V(1)$, which is the expected output voltage for a coincidence. Also, if we neglect $R_s + R_f$ compared with R, then if $m \neq n$, $v_0 = V(0) + V_\gamma$, and the output is clamped at a value V_γ above the $V(0)$ level.

excited inputs. The output increases by small steps as *m* increases from 0 to $n - 1$. This variation in level is known as logical noise. In an AND gate even the slight response (the noise) to something less than a complete coincidence is often undesirable. To reduce this effect a shunt diode *D* is added to Figure 1-95*b*, converting it to Figure 1-96, in order to clamp the output to a fixed voltage *V'* until all inputs are excited.

A CLAMPING DIODE SHUNTING THE OUTPUT. In the circuit of Figure 1-96 the 0 state $V(0)$ of the input is not the same as the 0 state *V'* of the output, and *V'* must be adjusted so that the individual diode currents $I_1, I_2, \ldots, I_n$ (which are nominally equal to one another) are each larger than I. For perfect diodes the restriction $I_1 > I$ means

$$I_1 = \frac{V' - V(0)}{R_s} > \frac{V_R - V'}{R} = I. \tag{1-128}$$

From (1-128) we must select $V' > V(0)$ in order to have a positive value for the current I_1. The diode *D* keeps the output close to *V'* for anything less than a complete coincidence but, of course, will not act as a perfect clamp because of the finite forward resistance R_f of *D*.

A DYNAMIC LOGIC SYSTEM. Consider the waveform at the output of the circuit of Figure 1-96. For simplicity, let us neglect the capacitances across the diodes. When the output pulse is formed, all diodes are reverse-biased and the output capacitance C_0 must charge through *R*. The output will therefore rise from *V'* to V_R with a time constant $T \equiv RC_0$. At the termination of the input pulses the series diodes conduct; hence they introduce a resistance $R' \equiv (R_s + R_f)/n$. The output capacitance now discharges with a time constant $T' \equiv RR'C_0/(R + R')$. Because $T' << T$, the trailing edge of the output pulse will decay much more rapidly than the leading edge will rise. The waveform when the peak value of the pulse exceeds V_R is shown in Figure 1-97*a*. If the peak value of the pulse is smaller than V_R, the rise time is improved, as shown in Figure 1-97*b*. If the diode capacitances are taken into account, the output waveforms are modified only slightly from those shown in Figure 1-97.

EMITTER-FOLLOWER LOGIC. The circuit of Figure 1-94 without modification functions as a positive AND gate. A negative AND gate is obtained by using *n-p-n* transistors.

Example 1-5

A radioactive source, in a certain mode of disintegration (decay), emits an alpha particle and a gamma ray in sequence, the intervening delay being entirely negligible. Explain how to determine the energy distribution of the alpha particles originating in this type of disintegration.

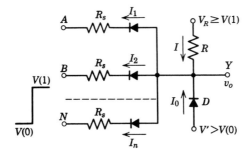

Fig. 1-96   A clamping diode D shunting the output of a positive AND gate.

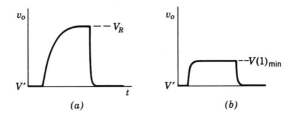

Fig. 1-97   Effect of output capacitance on output waveform in an AND circuit., (a) $V(1)_{min}$ > V_R; (b) $V(1)_{min}$ < V_R.

Solution   A coincidence counting system for doing this is shown in Figure 1-98. Each alpha particle is detected, when emitted in the appropriate direction, by a suitable counter (one of gas-filled, scintillation, or semiconductor radiation detectors). The alpha counter (detector 1) delivers a pulse whose amplitude is proportional to the energy of the particle incident on it.

The detectors are stimulated most frequently by uncorrelated particles. Such particles may originate in a different mode of decay from the source, or, perhaps, their partners did not happen to travel in the right direction to be intercepted by the other detector. The problem is to choose, from a large group of pulses in the two counters, just those which are correctly paired in time. This is the job of a coincidence gate. Evidently even the uncorrelated pulses occasionally occur almost simultaneously, simulating an event of the desired type. This background of random coincidences is unavoidable; its magnitude depends on the counting rate and on the resolving time of the gate. Although the true events have an inherent synchronization, the random ones do not; the gate must thus exercise time discrimination for cases of marginal overlap.

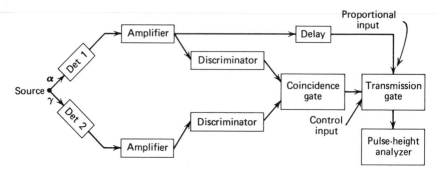

Fig. 1-98 Block diagram of coincidence counting system with transmission (sampling) gate for pulses from one channel.

The pulses from each detector are first amplified, as required. They are then applied to a discriminator. A discriminator is a circuit which performs the standardization of pulse amplitude while defining a suitable input threshold. Thus the discriminator in this coincidence system determines that the amplified pulses exceed a certain minimum threshold. The output from the discriminator is a standard pulse, both in amplitude and in width. The width determines, in large measure, the resolving time of the coincidence gate. Ideally, the gate would deliver an output whenever the input pulses overlapped at all; the resolving time in this case would be just the width of the pulses.

An output from the coincidence circuit signifies the occurrence of a desired event (or, occasionally, a random coincidence). This output is used to open a transmission gate, permitting the pulse from the alpha detector to pass while retaining its amplitude information (which is the analog of alpha-particle energy). Sorting and counting these transmitted pulses according to their amplitude then yields the desired energy spectrum. An instrument which performs this job is known as a pulse-height analyzer. The transmission gate must work with synchronized pulses to insure correct handling with minimum error. A delay (consisting of a calibrated length of cable in most cases) is usually required in the proportional pulse input line to make sure that the control pulse has had time to arrive. Notice that a transmission gate (also called a sampling gate) is distinct from the logic gate. A logic gate has to provide at the output a pulse or no pulse (or, alternatively, one voltage level or another), depending on the pulses (or voltages) present at the many gate inputs. A sampling gate usually has one signal input, and during the selected time interval, the output must reproduce faithfully the input waveform. Therefore, a sampling gate is also referred to as a linear gate.

The NOT Gate or Inverter

DEFINITION AND EXPRESSION.   The NOT gate has a single input and a single output and performs the operation of LOGIC NEGATION in accordance with the following definition: *the output of a NOT gate takes on the 1 state if and only if the input does not take on the 1 state.* The Boolean expression for negation is

$$Y = \text{NOT } A, \quad \text{or } Y = \overline{A}, \tag{1-129}$$

where the bar ($\overline{}$) indicates the NOT operation. The NOT gate is also called an *inverter*, since it inverts the sense of the output with respect to the input. When the input is at $V(0)$, the output must be at $V(1)$, and vice versa. Ideally, a NOT gate inverts a signal while preserving its shape and the binary levels between which the signal operates, as in Figure 1-99.

THE TRANSISTOR INVERTER.   Figure 1-100 shows a transistor inverter for positive logic having a 0 state of $V(0) = V_{EE}$ and a 1 state of $V(1) = V_{CC}$. When $v_i = V(0)$, Q is OFF, and so $v_0 = V_{CC} = V(1)$. On the other hand, when $v_i = V(I)$, Q is in saturation, and so $v_0 = V_{EE} = V(0)$ [neglecting the voltage V_{CE} (sat)]. The capacitor C across R_1 improves the transient response of the inverter, since this capacitor aids in the removal of the minority-carrier charge stored in the base when the signal changes abruptly between logic states.

In designing transistor inverters some transistor characteristics such as V_{EB}, h_{FE}, and I_{CBO} must particularly be taken into account.

THE TRANSFORMER INVERTER.   For pulse logic an inverting transformer can be used to perform the NOT operation. As shown in Figure 1-101, a dc restorer is added to establish the absolute levels of the two states at the output.

The Inhibitor or Anticoincidence Circuit

An *inhibitor* is a modified AND circuit in which a NOT gate precedes one terminal (N) of an AND gate. The inhibitor implements the following logical statements: *if $A = 1$, $B = 1$, ..., $M = 1$, then $Y = 1$, provided that $N = 0$. However, if $N = 1$, then the coincidence of $A, B, ..., M$ is inhibited, and $Y = 0$.* The inhibitor is also called an *anticoincidence* circuit. The Boolean expression for inhibitor is

$$Y = A \text{ and } B \text{ and } \ldots \text{ and } M \text{ and not } N$$

or

$$Y = AB \ldots M\overline{N}. \tag{1-130}$$

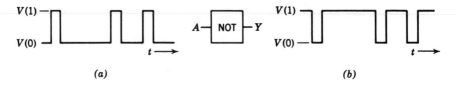

(a)   (b)

Fig. 1-99   (a) Input A and (b) output Y of a NOT gate.

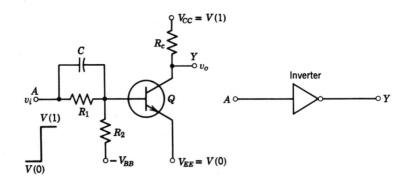

Fig. 1-100   A NOT circuit for positive logic. A typical example is min value of h_{FE} of Q = 30, $V_{CC} = V(1) = 12$ volts, $-V_{BB} = -12$ volts, $V_{EE} = 0$ V (ground potential), $R_c = 2.2$ K, $R_1 = 15$ K, $R_2 = 100$ K, $C = 100$ pF, $v_0 = 12$ volts for $v_i = 0$, and $v_0 = 0$ for $v_i = 12$ volts. (A similar circuit using a *p-n-p* transistor is employed for a negative-logic NOT circuit.)

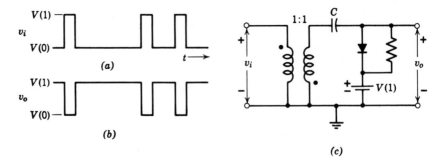

Fig. 1-101   Input (a) and output (b) of the transformer inverter circuit (c).

112

Figure 1-102 shows a positive-logic AND gate with a negation input terminal. This circuit is an inhibitor satisfying the logic given in the truth table of Figure 1-103. Considering the corresponding condition, an appropriate argument can be made to verify each line of the truth table.

The EXCLUSIVE OR Gate

An EXCLUSIVE OR gate has two inputs and a single output, and it

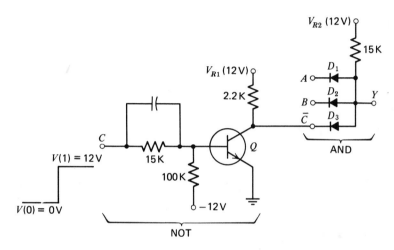

Fig. 1-102   A combination of the AND gate of Figure 1-95*b* and the NOT gate of Figure 1-100 with emitter grounded.

Input			Output
A	*B*	*C*	*Y*
1	0	0	0	0

	A	*B*	*C*	*Y*
1	0	0	0	0
2	0	1	0	0
3	1	0	0	0
4	1	1	0	1
5	0	0	1	0
6	0	1	1	0
7	1	0	1	0
8	1	1	1	0

Figure 1-103   Truth table for $Y = AB\bar{C}$.

operates in accordance with the following definition: *the output of a two-input EXCLUSIVE OR assumes the 1 state if one and only one input assumes the 1 state.* The above definition is equivalent to the statement: "if $A = 1$ or $B = 1$ but not simultaneously, then $Y = 1$." The Boolean expression is

$$Y = (A + B)\,(\overline{AB}). \tag{1-131}$$

This function is implemented in logic diagram form in Figure 1-104a, where the first A block represents an AND gate and the second A block with a small circle represents an AND with a negation (NOT) input. This negation before the second AND could be put equally well at the output of the first AND without changing the logic.

A second logical statement equivalent to the definition of the EXCLUSIVE OR (OE) is the following: "if $A = 1$ and $B = 0$, or if $B = 1$ and $A = 0$, then $Y = 1$." In Boolean notation

$$Y = A\overline{B} + B\overline{A}. \tag{1-132}$$

The block diagram which satisfies this logic is shown in Figure 1-104b.

The NAND and NOR Gates

A NAND gate is a NOT-AND or negated AND circuit. The Boolean expression is $Y = \overline{AB}$. The truth table is given in Figure 1-105. A single input NAND is a NOT. A NAND followed by a NOT is an AND. If the transistor NOT gate is placed after the diode AND in Figure 1-102, then we obtain the NAND circuit in Figure 1-106. Consider the NAND operation. If any input is at 0V, the junction point P of the diodes is at 0V, since a diode conducts and clamps this point to $V(0) = 0$. The base voltage of the transistor Q is then $V_B = -(12)(15/115) = -1.56V$. Therefore Q is cut off and Y is at 12 V, or $Y = 1$. This result verifies the first three lines of the truth table of Figure 1-105. If all inputs are at $V(1) = 12V$, then all diodes are reverse-biased. Suppose in this case Q is in saturation; then with $V_{BE} = 0$ the voltage at P is $(12)(15/30) = 6V$.

Therefore with 12 V at each input all diodes are reverse-biased by 6 V, so that they are nonconducting. Then we find $I_B = 12/30 - 12/100 = 0.28$ mA, $I_C = 12/2.2 = 5.45$ mA, and $(h_{FE})_{min} = 5.45/0.28 = 19$. Hence Q will indeed be in saturation if $h_{FE} \geq 19$. Under these conditions the output is at 0V or $Y = 0$. This result verifies the last line of the truth table.

Now, consider the characteristic of the output circuit of Figure 1-106. Neglecting the inherent speed limitations of the transistor, the rise time of the output, when Q is cut off, depends on the shunt capacitance C_s and the collector resistance R_c. If an increased value of V_{CC} $[> V(1)]$ is

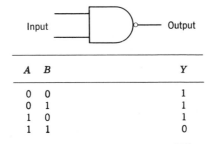

$$Y = (A + B)(\overline{AB})$$

(a)

$$Y = A\overline{B} + B\overline{A}$$

(b)

Fig. 1-104   Two logic blocks for the EXCLUSIVE OR (OE) gate.

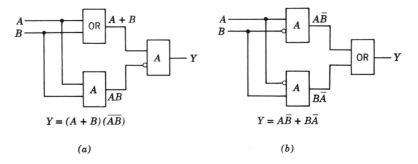

A	B	Y
0	0	1
0	1	1
1	0	1
1	1	0

Fig. 1-105   Truth Table for $Y = \overline{AB}$.

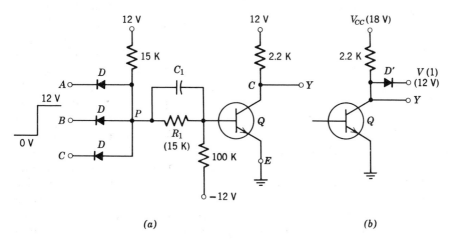

(a)

(b)

Fig. 1-106   (a) A three-input positive NAND (or negative NOR) gate; (b) a collector clamping diode D' for improving characteristics of the output circuit.

115

used, then for fixed values of C_s and R_c, less time is required to reach the particular voltage at which the next stage is driven into saturation. In this case, a collector clamping diode for limiting V_{CC} to $V(1)$ is often added, as in Figure 1-106*b*. Thus this diode not only prevents the following stage from being driven too heavily into saturation, but also helps to reduce the time necessary for discharging C_s, when Q is driven into saturation, by providing a lower collector starting voltage.

A NOR gate is a NOT-OR gate, that is, a negation following an OR. The Boolean expression is $Y = \overline{A + B}$. A practical form of positive NOR gate is shown in Figure 1-107. It is implemented by a cascade of a diode OR and a transistor inverter (NOT). The specific binary levels are $V(0) = 0$ and $V(1) = 12$ V. The $-V_{BB} = -12$ V is also used as the reference voltage V_R for the diode OR, and hence the diode output resistor R (see Figure 1-91) is omitted. Now, let us consider the NOR operation. If all inputs are in the 0 state, all diodes conduct and the input to the inverter is 0 V. If any input is high, the diode connected to this input conducts and all other diodes are reverse-biased. The voltage at the diode node P is now $V(1) = 12$ V. Therefore, from input to point P the OR function has been satisfied. From P to the output we have an inverter exactly like that in Figure 1-100, which was discussed previously.

External Characteristics of the flip-flop

A flip-flop consists of two NOT circuits interconnected in the manner shown in Figure 1-108*a*. Each NOT might be, for example, the transistor inverter of Figure 1-100. The most important characteristic of the flip-flop is that, because of the interconnection, the circuit can persist indefinitely in a state in which one device (say Q_1) is ON while the other (Q_2) is OFF. A second stable state of the flip-flop is one in which the roles of the two devices are interchanged so that Q_1 is OFF and Q_2 is ON. Because the flip-flop has two stable states it may be used to store one bit of information. For these reasons the flip-flop is also known as a *binary*.

In Figure 1-108 the output Y may be taken from one collector and the output $\overline{Y}$ from the other collector. Then the flip-flop has two stable states, one in which $Y = 1$ and $\overline{Y} = 0$ and the other in which $Y = 0$ and $\overline{Y} = 1$. An excitation of the set S input causes the flip-flop to establish itself in the state $Y = 1$. If the binary is already in that state the excitation has no effect. A signal at the reset R input causes the flip-flop to establish itself in the state $Y = 0$. If the binary is already in that state, the excitation has no effect. A triggering signal applied to the T (trigger) input causes the flip-flop to change its state regardless of the existing state of the binary. Thus each successive triggering signal applied to T induces a transition. This type of excitation is known as symmetrical triggering and is used in binary counters, etc. Unsymmetrical triggering through the S or R input

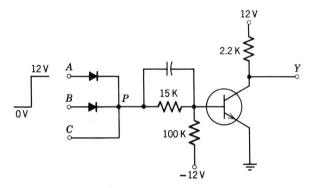

Fig. 1-107   A practical form of positive NOR (or negative NAND) gate. The direct connection C is convenient for expanding the number of inputs by adding more diodes as needed.

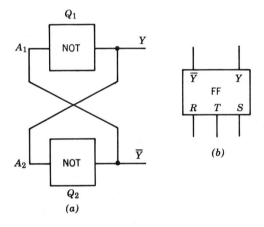

Fig. 1-108   (a) A FLIP-FLOP consisting of two NOT circuits; (b) the logic symbol. An input to T (trigger) effectively applies excitation to S (set) and R (reset) simultaneously.

is most useful in logic applications. In unsymmetrical triggering the triggering signal is effective in inducing a transition in only one direction. A second triggering signal from a separate source must be introduced in a different manner to achieve the reverse transition.

1-18 TRANSISTOR OSCILLATORS

Sinusoidal Oscillators

Transistor oscillator circuits are similar in many respects to amplifiers except that a portion of the output power is returned to the input network in phase with the starting power (regenerative or positive feedback) to sustain oscillation. Dc bias-voltage requirements for oscillators are similar to those for amplifiers. The maximum frequency of oscillation of a transistor is defined as the frequency at which the power gain is unity. Because some power gain is required in an oscillator circuit to overcome losses in the feedback network, the operating frequency must be some value below the transistor maximum frequency of oscillation. The frequency-determining elements of an oscillator circuit may consist of an inductance-capacitance (*LC*) network, a crystal, or a resistance-capacitance (*RC*) network. An *LC* tuned circuit may be placed in either the base circuit or the collector circuit of a common-emitter transistor oscillator.

A quartz crystal is often used as the frequency-determining element in a transistor oscillator circuit because of its extremely high Q (narrow bandwidth) and good frequency stability over a given temperature range. A quartz crystal can be operated as either a series or parallel resonant circuit. As indicated in Figure 1-109, the electrical equivalent of the mechanical vibrating characteristic of the crystal can be represented by a resistance R, an inductance L, and a capacitance C_s in series. The lowest impedance of the crystal occurs at f_s, the series-resonance frequency of C_s and L; f_s is then determined only by the mechanical vibrating characteristics of the crystal. At frequencies above the series-resonant frequency, the combination of L and C_s has the effect of a net inductance because the inductive reactance of L is greater than the capacitive reactance of C_s. This net inductance forms a parallel resonant circuit with C_p (the electrostatic capacitance between the crystal electrodes) and any circuit capacitance across the crystal. The impedance of the crystal is highest at f_p, the parallel resonant frequency; f_p is then determined by both the crystal and the externally connected circuit elements.

A typical crystal oscillator is shown in Figure 1-110. Here Q_1 is an amplifier with a positive feedback from collector to emitter, with the 100 kHz crystal as feedback element. Q_2 is included to match the high-impedance collector circuit of Q_1 to the low-impedance emitter.

Nonsinusoidal Oscillators

Oscillator circuits which produce nonsinusoidal output waveforms use a regenerative circuit in conjunction with resistance-capacitance (*RC*) or resistance-inductance (*RL*) components to produce a switching action. The charge and discharge times of the reactive elements (*RC* or *L/R*) are used to produce saw-tooth, square, or pulse-output waveforms.

MULTIVIBRATORS.   A multivibrator is essentially a nonsinusoidal two-

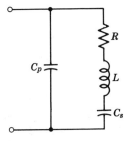

Fig. 1-109 Equivalent circuit of quartz crystal.

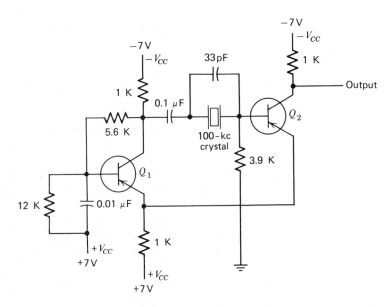

Fig. 1-110 Typical crystal oscillator. (From the manual *Operating and Maintenance Instructions for* 400 *Channel Analyzer System Model* 20631/2, Radiation Counter Laboratories, Inc.)

stage oscillator in which one stage conducts while the other is cut off until a point is reached at which the conditions of the stages are reversed. This type of oscillator is normally used to produce a square-wave output. In the *RC*-coupled common-emitter multivibrator of Figure 1-111, the output of Q_1 is coupled to the base of Q_2 through the feedback capacitor C_1, and the output of Q_2 is coupled to the base of Q_1 through the feedback capacitor C_2. An increase in the collector current

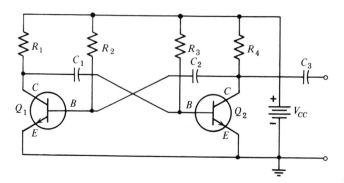

Fig. 1-111   The free-running collector-coupled astable multivibrator.

of Q_1 causes a decrease in the collector voltage which, when coupled through C_1 to the base of Q_2, causes a decrease in the collector current of Q_2. The resultant rising voltage at the collector of Q_2, when coupled through C_2 to the base of Q_1, drives Q_1 further into conduction. This regenerative process occurs rapidly, driving Q_1 into heavy saturation and Q_2 into cutoff. Q_2 is maintained in a cutoff condition by C_1 (which was previously charged to the supply voltage through R_1) until C_1 discharges through R_3 toward the collector-supply potential. When the junction of C_1 and R_3 reaches a slight positive voltage, however, Q_2 begins to start into conduction and the regenerative process reverses. Q_2 then reaches a saturation condition, Q_1 is cut off by the reverse bias applied to its base through C_2, and the junction of C_2 and R_2 starts charging toward the collector-supply voltage. The two transistors are therefore alternately cut off. The result is a square-wave oscillation, the frequency of which is determined by the time constants of the components.

The multivibrator described above oscillates regularly between the two states, and is referred to as an astable multivibrator. There are two other types of multivibrators, namely the monostable and the bistable circuits. A monostable circuit has one stable state in either of the stable regions (cutoff or saturation); an external pulse triggers the transistor to the other stable region, but the circuit then switches back to its original stable state after a period of time determined by the time constants of the circuit elements. A bistable (flip-flop) circuit has a stable state in each of the two stable regions. The transistor is triggered from one stable state to the other by an external pulse, and a second trigger pulse is required to switch the circuit back to its original stable state.

BLOCKING OSCILLATORS. Another form of nonsinusoidal oscillator is the blocking oscillator which conducts for a short period of time and is blocked (cut off) for a much longer period. A basic blocking oscillator is shown in Figure 1-112. The base coil (winding 1-2 of transformer T_1) and collector coil (winding 3-4) must be connected for regenerative feedback. This feedback through winding 1-2 and capacitor C causes current through the transistor to rise rapidly until saturation is reached. The transistor is then cut off until C discharges through resistor R. The output waveform is a pulse, the width of which is primarily determined by winding 1-2. The time between pulses (blocking time) is determined by the time constant RC.

A triggered blocking oscillator (emitter-timing) and the equivalent circuit are shown in Figure 1-113. Applying Kirchhoff's voltage law to the outside loop (Figure 1-113b), encompassing both the collector and base meshes, gives

$$V = \frac{V_{CC}}{n+1},\qquad(1\text{-}133)$$

where V is the voltage drop across the collector winding during the pulse. Since the voltage drop across R is

$$V_{EN} = nV = (i_C + i_B)\,R,$$

the emitter current which is constant is given by

$$-i_E = i_C + i_B = \frac{nV}{R} = \frac{n}{n+1}\frac{V_{CC}}{R}.\qquad(1\text{-}134)$$

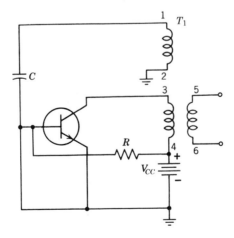

Fig. 1-112 Basic transistor-blocking oscillator.

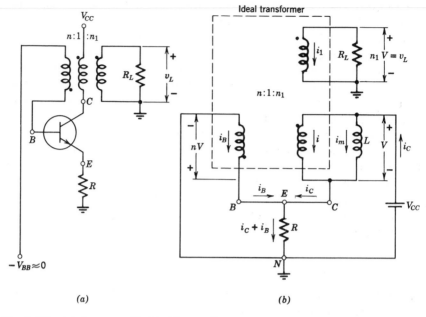

Fig. 1-113 (a) A monostable blocking oscillator with emitter timing; (b) the equivalent circuit from which to calculate the current and voltage waveforms.

Since the sum of the ampere turns in the ideal transformer is zero,

$$i - ni_B + n_1 i_1 = 0. \tag{1-135}$$

The current in the load circuit is

$$i_1 = -\frac{n_1 V}{R_L}. \tag{1-136}$$

Since V is a constant, the magnetizing current is given by

$$i_m = \frac{Vt}{L}. \tag{1-137}$$

From Kirchhoff's current law at the collector node, we have

$$i = i_C - i_m = i_C - \frac{Vt}{L}. \tag{1-138}$$

Substituting from (1-136) and (1-138) into (1-135), we have

$$i_C - \frac{Vt}{L} - ni_B - \frac{n_1^2 V}{R_L} = 0. \tag{1-139}$$

Solving (1-134) and (1-139) and using (1-133), we obtain

$$i_C = \frac{V_{CC}}{(n+1)^2} \left(\frac{n^2}{R} + \frac{n_1^2}{R_L} + \frac{t}{L} \right) \tag{1-140}$$

and

$$i_B = \frac{V_{CC}}{(n+1)^2} \left(\frac{n}{R} - \frac{n_1^2}{R_L} - \frac{t}{L} \right). \tag{1-141}$$

Notice that the collector-current waveform is trapezoidal with a positive slope, the base current is also trapezoidal with a negative slope, and the emitter current is constant during the pulse. These current waveforms and the voltage waveforms are pictured in Figure 1-114.

At $t = 0 +$, $i_C < h_{FE} i_B$, the operating point on the collector characteristics of Figure 1-115 is at point P and the transistor is in saturation. As time passes, i_C increases and the operating point moves up the saturation line in Figure 1-115. While i_C grows with time, i_B is decreasing and eventually point P' is reached at $t = t_p$, where $i_B = I_B$ and

$$i_C = h_{FE} I_B. \tag{1-142}$$

At this point P' the transistor comes out of saturation and enters its active region. Because the loop gain exceeds unity in the active region, the transistor is quickly driven to cutoff by regenerative action, and the

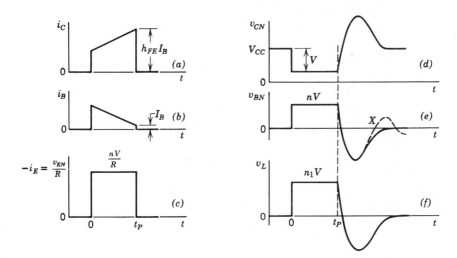

Fig. 1-114   The current and voltage waveforms in a monostable blocking oscillator with emitter timing (Figure 1-113). If the damping is inadequate, the backswing may oscillate, as indicated by the dashed curve in (e), and regeneration will start again at the point marked X.

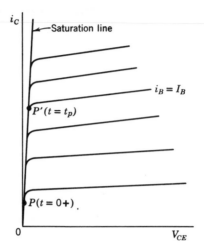

Fig. 1-115   Collector characteristics. The path of the collector current is along the saturation line from P to P'. The pulse ends at P', where the transistor comes out of saturation.

pulse ends. Since the regeneration which terminates the pulse starts when the transistor comes out of saturation, the pulse width[1] t_p is determined by the condition given by (1-142).

At the termination of the pulse there exists a current i_m in the magnetizing inductance of the transformer. Since the current through an inductor cannot change instantaneously, i_m must continue to flow even after $t = t_p$, when the transistor currents have dropped to zero. Since the capacitance of the transformer is small, i_m decays rapidly and hence induces large voltage overshoots at the collector, base, and load, as in Figure 1-114. These overshoots must not be so large as to exceed the breakdown voltage BV_{CE} or BV_{EB}. It should be noted that adequate damping of the backswing which occurs at the termination of the pulse is absolutely essential to the operation of the blocking oscillator. In order to suppress the transformer oscillations without loading the blocking oscillator during the pulse interval, a damping resistor R' in series with a diode may be shunted across the transformer. With an n-p-n transistor, as in Figure 1-113, the anode of the diode is at the collector side and the cathode at the supply-voltage side, so that the diode does not conduct

---

[1] Applying (1-142) to (1-140) and (1-141), we obtain $t_p = (nL/R) [(h_{FE} - n)/(h_{FE} + 1)] - (n_1{}^2L/R_L)$. (1-143a) Since usually $\tfrac{1}{5} \leq n \leq 1$, then $h_{FE} >> n$ and (1-143a) is approximated by $t_p \approx (nL/R) - (n_1{}^2L/R_L)$. (1-143b) If the second term in (1-143b) exceeds the first term, then t_p is negative, a situation which is impossible. In order for the loop gain to exceed unity, the following inequality must be valid: $R_L > (n_1{}^2R/n) [(h_{FE} + 1)/(h_{FE} - n)]$.

during the pulse interval but does conduct during the overshoot. The resistor R' is selected to be smaller than the critical damping resistance.

There are applications in which the blocking oscillator must drive a variable load, for example, a ferrite-core memory where the load depends upon the number of cores to be excited. A method of obtaining a pulse duration which is independent of the loading involves placing R_L in the collector leg of a second transistor Q_2 whose base current is the collector current of the blocking-oscillator transistor Q_1, as in Figure 1-116.

1-19 FLIP-FLOP CIRCUITS

The Basic Bistable Circuit

Bistable circuits usually employ two-stage regenerative amplifiers and they are known by a wide variety of names such as bistable multivibrator, scale-of-2 toggle circuit, trigger circuit, flip-flop, and binary. They are used for the performance of many digital operations such as counting and storing of binary information. Their principal characteristics are:

1. They have two stable operating states.
2. The regenerative action carries the circuit from one state to another, and this action is essentially independent of the amplitude of the trigger pulse (provided that the trigger pulse has reached a certain critical minimum amplitude) except in high-speed operation.

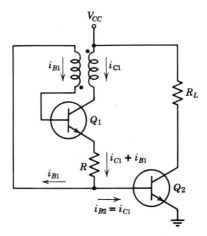

Fig. 1-116 The load R_L is switched ON and OFF through Q_2, which is driven by the blocking oscillator collector current. The effective supply voltage of Q_1 is $V_{CC} - V_{BE2}$ (sat) $\simeq V_{CC}$. The load current is $[V_{CC} - V_{CE2}$ (sat)$]/R_L \simeq V_{CC}/R_L$ for a time $t_p \simeq nL/R$ and is zero outside this interval.

3. The circuit remains in a stable state for an arbitrarily long time when no trigger signal is applied, and this state is determined by the preceding pulse.

This last condition is a property associated with "memory," and is, in effect, a condition required of all bistable devices and circuits.

The basic flip-flop shown in Figure 1-117 is the Eccles-Jordan type multivibrator circuit. The resistive and bias values of this circuit are chosen so that the initial application of dc power causes one transistor to be cut off and the other to be driven into saturation. Because of the feedback arrangement, each transistor is held in its original state by the condition of the other. The application of a positive trigger pulse to the base of the OFF transistor or a negative pulse to the base of the ON transistor switches the conducting state of the circuit. The new condition is then maintained until a second pulse triggers the circuit back to the original condition. Thus a trigger pulse at input A will change the state of the circuit. An input of the same polarity at input B or an input of opposite polarity at input A will then return the circuit to its original state. (Collector triggering can be accomplished in a similar manner.) The speed-up capacitors C_3 and C_4 are used to help remove the stored base charge at the moment at which the saturated ON transistor is to be turned off, so that the regenerative switching action is speeded up. The output of the circuit is a unit step voltage when one trigger is applied, or a square wave when continuous pulsing of the input is used.

A Self-biased Flip-flop

A self-biased flip-flop using p-n-p transistors is shown in Figure 1-118. The common emitter resistor R_e is used to provide self-bias instead of the fixed bias applied to the base-emitter circuit. A capacitor (not shown) is often used to bypass R_e in order to keep the self-bias almost constant during the transition time. A maximum of V_{CE} will appear across the transistor that is OFF. The condition that the supply voltage be reasonably smaller than the transistor-collector breakdown voltage BV_{CE} will usually restrict $-V_{CC}$ to the order of several tens of volts. Under saturation conditions the collector current I_C is a maximum. Therefore, R_c must be selected so that the value of I_C does not exceed the maximum permissible value. The values of R_1, R_2, and the self-bias must be chosen so that in one state the base current is large enough to drive the transistor into saturation and in the second state the emitter junction must be below cutoff. The signal at a collector, called the output swing, is the change in collector voltage resulting from a transition from one state to the other.

When the flip-flop is used to drive other circuits, there are shunting

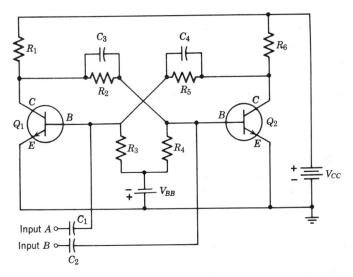

Fig. 1-117 A basic fixed-bias transistor binary—Eccles-Jordan-type bistable multi-vibrator. (A typical example: $V_{CC} = 12$ V, $V_{BB} = -12$ V, $R_1 = R_6 = 2.2$ K, $R_2 = R_5 = 15$ K, $R_3 = R_4 = 100$ K, $C_3 = C_4 = 220$ pF.)

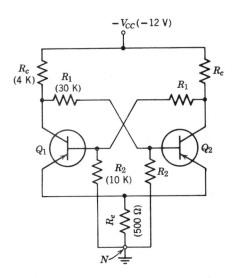

Fig. 1-118 A self-biased flip-flop using p-n-p transistors. A capacitor is often used to by-pass R_e.

127

loads at one or both collectors. These loads reduce the magnitude of V_{C1} of the OFF transistor, and so decrease the output swing as well as the base current I_{B2}. Because of the reduced I_{B2}, it is possible that Q_2 may not be driven into saturation. Hence the flip-flop circuit components must be chosen so that under the heaviest load which the flip-flop drives, one transistor will remain in saturation while the other will be cut off.

Example 1-6

The self-biased flip-flop of Figure 1-118 uses *p-n-p* germanium transistors. The connections between the base of Q_1 and the collector of Q_2 are shown in Figure 1-119*a*, whereas the connections from the collector of Q_1 to the base of Q_2 are given in Figure 1-119*b*. Calculate the stable-state currents and voltages and find the minimum value of h_{FE} which will keep the ON transistor in saturation.

Solution   Assume that Q_1 is OFF and Q_2 is ON (in saturation). The equivalent circuit for Q_2 is shown in Figure 1-120, where we have replaced the collector circuit of Q_2 in Figure 1-119*a* by its Thevenin voltage

$$\frac{-V_{CC}(R_1 + R_2)}{R_1 + R_2 + R_c} = \frac{(-12)(30 + 10)}{30 + 10 + 4} = -10.9 \text{ V}$$

and its Thevenin resistance

$$\frac{R_c(R_1 + R_2)}{R_1 + R_2 + R_c} = \frac{(4)(40)}{44} = 3.64 \text{ K}.$$

Likewise, the Thevenin equivalent of the base circuit of Q_2 is obtained from Figure 1-119*b* as a voltage

$$\frac{-V_{CC} R_2}{R_1 + R_2 + R_c} = \frac{(-12)(10)}{44} = -2.73 \text{ V}$$

in series with a resistance

$$\frac{R_2(R_1 + R_2)}{R_1 + R_2 + R_c} = \frac{(10)(34)}{44} = 7.73 \text{ K}.$$

Assume that V_{BE} (sat) ≈ -0.3 V and V_{CE} (sat) ≈ -0.1 V (see Table 1-5). The equations of Kirchhoff's voltage law are

$$2.73 - 0.3 + I_{B2}(7.73 + 0.5) + I_{C2}(0.5) = 0$$

and

$$10.9 - 0.1 + I_{B2}(0.5) + I_{C2}(3.64 + 0.5) = 0.$$

Solving, we obtain $I_{C2} = -2.59$ mA and $I_{B2} = -0.138$ mA. Therefore

$$(h_{FE})_{min} = \frac{I_{C2}}{I_{B2}} = \frac{-2.59}{-0.138} = 18.8.$$

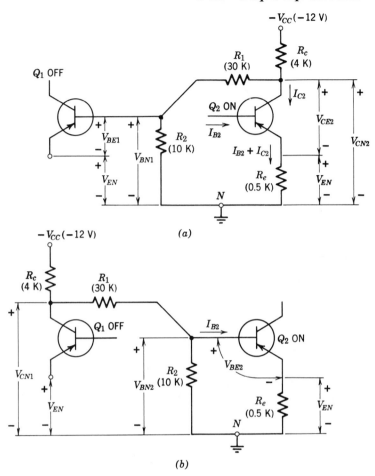

Fig. 1-119 Circuit of Figure 1-118 redrawn to indicate the connections (*a*) between the base of Q_1 and the collector of Q_2 and (*b*) from the collector of Q_1 to the base of Q_2.

From Figures 1-119*a* and *b* we find

$$V_{EN} = (I_{B2} + I_{C2}) R_e = (-0.138 - 2.59) (0.5) = -1.36 \, \text{V},$$
$$V_{CN2} = V_{CE2} + V_{EN} = -0.1 - 1.36 = -1.46 \, \text{V},$$
$$V_{BN2} = V_{BE2} + V_{EN} = -0.3 - 1.36 = -1.66 \, \text{V},$$
$$V_{BN1} = V_{CN2} \frac{R_2}{R_1 + R_2} = (-1.46) \left(\frac{10}{40} \right) = -0.37 \, \text{V},$$
$$V_{BE1} = V_{BN1} - V_{EN} = -0.37 + 1.36 = +0.99 \, \text{V}.$$

This last value indicates that Q_1 is certainly OFF, since only about $+0.1$ V

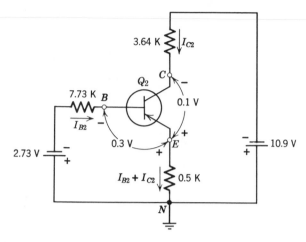

Fig. 1-120   Equivalent circuit when Q_2 in Figure 1-118 is in saturation.

of V_{BE} is required to cut off a *p-n-p* germanium transistor (see Table 1-5). From Figure 1-119*b*,

$$V_{CN1} = \frac{-V_{CC}\,R_1}{R_c + R_1} + \frac{V_{BN2}\,R_c}{R_c + R_1} = \frac{(-12)\,(30)}{34} + \frac{(-1.66)\,(4)}{34} = -10.8 \text{ V}.$$

The output swing is $V_\omega = V_{CN2} - V_{CN1} = -1.46 + 10.8 = 9.3$ V. In summary, we have

I_{C1}	$= 0$ mA	I_{C2}	$= -2.59$ mA
I_{B1}	$= 0$ mA	I_{B2}	$= -0.14$ mA
V_{CN1}	$= -10.8$ V	V_{CN2}	$= -1.46$ V
V_{BN1}	$= -0.37$ V	V_{BN2}	$= -1.66$ V
V_{EN}	$= -1.36$ V	V_ω	$= 9.3$ V
$(h_{FE})_{min}$	$= 18.8$		

Symmetrical Triggering

For counter applications, it is necessary for pulses at a single input to make the two sides of the flip-flop conduct alternately. Outputs from the flip-flop must have characteristics suitable for triggering other similar flip-flops. When the counting period is finished, it is generally necessary to reset the counter by a trigger pulse to one side of all flip-flops simultaneously. Shift registers and ring counters have similar triggering requirements.

In applying a trigger to one side of a flip-flop, it is preferable to have the trigger turn a transistor off rather than on. The OFF transistor usually has a reverse-biased emitter junction. This bias voltage must be overcome by the trigger before switching can start. Furthermore, some transistors have slow turn on characteristics, which result in a delay between the application of the trigger pulse and the actual switching. On the other hand, since no bias has to be overcome, there is less delay in turning off a transistor. As turn-off begins, the flip-flop itself turns the other side on.

A lower limit on trigger power requirements can be determined by calculating the base charge required to maintain the collector current in the ON transistor. The trigger source must be capable of neutralizing this charge in order to turn off the transistor. It has been determined that the base charge for a nonsaturated transistor is approximately $Q_B = \tau_c I_c$, where τ_c is the collector time constant. This indicates that circuits utilizing high-speed transistors at low collector currents will require the least trigger power. If the ON transistor was in saturation, the trigger power must also include the stored base charge. This so-called stored charge is a measure of the amount of charge which exists in the base region of the transistor at the time that forward bias is removed. This stored charge supports an undiminished collector current in the saturation region for some finite time before complete switching is effected. This delay interval, called the "storage time," depends on the degree of saturation into which the transistor is driven. The base stored charge is given by $Q_s = \tau_b I_{BX}$, where τ_b is the effective lifetime in the saturated region, and I_{BX} is the current which is permitted to flow into the base in excess of that required to saturate the transistor.

In designing counters, shift registers, or ring counters, it is necessary to make alternate sides of a flip-flop conduct on alternate trigger pulses. There are so-called steering circuits which accomplish this. At low speeds, the trigger may be applied at the emitters, as indicated in Figure 1-121. For reliable operation it is important that the trigger pulse be shorter than the cross-coupling time constant. The limitation of this circuit lies in the high trigger current required. The effect of trigger-pulse repetition rate can be analyzed. In order that each trigger pulse produce reliable triggering, it must find the circuit in exactly the same state as the previous pulse found it. This means that all the capacitors in the circuit must stop charging before a trigger pulse is applied. If they do not, the result is equivalent to reducing the trigger-pulse amplitude. The transistor being turned off presents a low impedance, permitting the trigger capacitor C_T to charge rapidly. The capacitor must then recover its initial charge through another impedance which is generally much higher. The recovery time constant can limit the maximum pulse rate.

Steering circuits using diodes are shown in Figures 1-122 and 1-123.

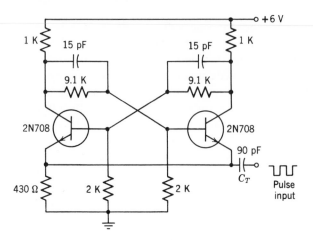

Fig. 1-121 An example of emitter triggering. Maximum trigger rate exceeds 2 MHz with trigger amplitude from 4 to 12 V. (From *Transistor Manual* 7th ed., General Electric Company, 1964.)

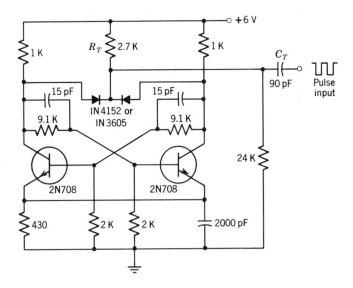

Fig. 1-122 An example of collector triggering. Maximum trigger rate exceeds 5 MHz with trigger amplitude from 4 to 12 V. (From *Transistor Manual*, 7th ed., General Electric Company, 1964.)

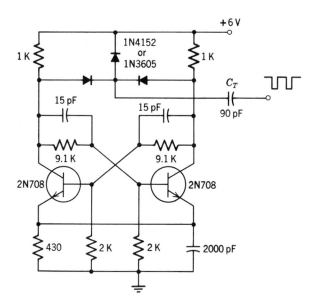

Fig. 1-123   Another example of collector triggering. Diode to supply voltage reduces trigger power and extends maximum trigger rate. (From *Transistor Manual*, 7th ed., General Electric Company, 1964.)

In Figure 1-122 the collectors are triggered by applying a negative pulse. As a diode conducts during triggering, the trigger pulse is loaded by the collector load resistance. When triggering is accomplished, the capacitor recovers through the biasing resistor R_T. To minimize trigger loading, R_T should be large; to aid recovery, it should be small. To avoid the recovery problem mentioned above, R_T can be replaced by a diode as in Figure 1-123. The diode's low forward impedance ensures fast recovery, and its high back impedance avoids shunting the trigger pulse during the triggering period. Collector triggering requires a relatively large-amplitude low-impedance pulse but has the advantage that the trigger pulse adds to the switching collector waveform to enhance the speed. Large variations in trigger-pulse amplitude are also permitted.

The steering circuit shown in Figure 1-124 is an example of base triggering. The principal differences between collector triggering and base triggering are quantitative; the latter requires less trigger energy but a more accurately controlled trigger amplitude.

Unsymmetrical Triggering

Unsymmetrical triggering, using two triggering sources, finds exten-

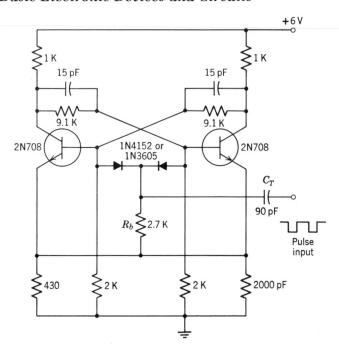

Fig. 1-124 An example of base triggering. Maximum trigger rate exceeds 5 MHz with trigger amplitude from 0.75 to 2 V. A diode can replace R_b to shorten the recovery time. (From *Transistor Manual*, 7th ed., General Electric Company, 1964.)

sive application in logic circuitry. A method of triggering a binary which allows the flip-flop to respond to only one polarity of pulse is given in Figure 1-125. When Q_1 is OFF the drop across the diode D is zero. D will transmit a negative step or pulse to the base of Q_2, thus turning Q_2 OFF. If the triggering rate is high, it may be necessary to replace the resistor R with a diode. The placement of this diode in the circuit is such that it is reverse-biased during the pulse; however it does conduct after the pulse in order to quickly remove the charge accumulated on C.

We may use positive or negative logic to describe the operation of the flip-flop of Figure 1-125. For a positive logic system the output Y is taken from the collector of the *n-p-n* transistor Q_2 and the output $\overline{Y}$ is taken from the collector of Q_1. After a triggering signal (negative pulse) is applied at the set terminal S, Q_2 will be OFF and Q_1 ON, so that $Y = 1$ and $\overline{Y} = 0$. Likewise, after a triggering signal is applied at the reset terminal R, Q_2 will be ON and Q_1 OFF, so that $Y = 0$ and $\overline{Y} = 1$.

Another method for triggering a flip-flop unsymmetrically is shown in

Figure 1-126. In this circuit a positive input at S (set) will forward-bias $CR6$, and turn Q_1 OFF; turning Q_1 OFF produces a negative voltage on the base of Q_2, which turns Q_2 ON. A positive input at R (reset) will forward-bias $CR7$, and turn Q_2 OFF; turning Q_2 OFF produces a negative voltage on the base of Q_1, which turns Q_1 ON. The diode $CR4$ ($CR9$) is reverse-biased during the negative output pulse from $Q_1(Q_2)$, but it does conduct when a positive pulse appears at the collector. The diode $CR5$ ($CR8$) is connected in the reverse direction across the emitter-base junction, so that the turnoff spike is clipped. This prevents Q_1 (Q_2) from local breakdown in the miniature regions of the junction.

A Four-binary Counting Chain

A cascade of flip-flops (Figure 1-127) is used as a counting chain. The outputs Y and $\overline{Y}$ are taken from the collectors. The trigger input terminal (T) is a point at which the flip-flop may be triggered symmetrically, so that each successive triggering signal will reverse the state of the binary. For the reason mentioned above, it is preferable to have the applied trigger turn a transistor off rather than on. Hence for flip-flops constructed of *n-p-n* transistors, as in Figure 1-127, a negative pulse is required for triggering. Any of the flip-flops receiving a positive pulse will not respond by making a transition.

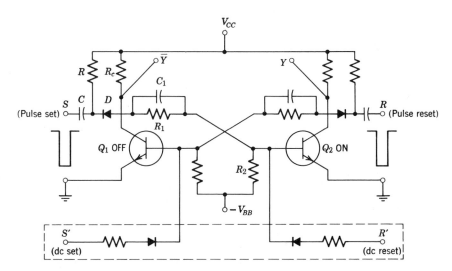

Fig. 1-125   A method for triggering a flip-flop unsymmetrically. (The dashed box indicates the dc triggering arrangement.)

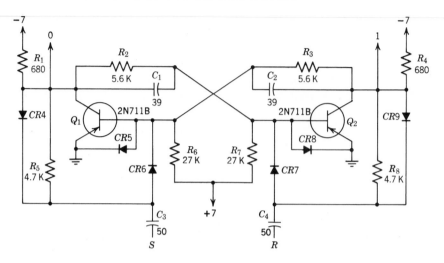

Fig. 1-126 Another method for triggering a flip-flop unsymmetrically. (From the man-
~~ual~~ *Operating and Maintenance Instructions for 400 Channel Analyzer System*, Model
20631/2, Radiation Counter Laboratories, Inc.)

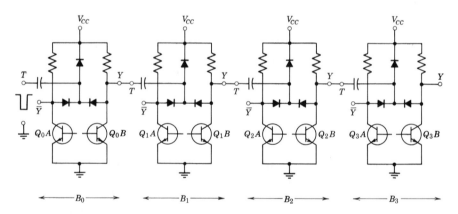

Fig. 1-127 Circuitry of a four-binary counting chain. The cross-coupling connections
from collectors to bases have been omitted. (For each complete binary circuit, see Figure
1-123.)

Let us use positive logic to describe our systems; we thus consider
that a flip-flop is in state 0 when the *n-p-n* transistor whose collector is
connected to the Y output conducts. The flip-flop is in state 1 when this
device is not conducting. When the flip-flop is in the state 0, the voltage at
Y will be low, and when the flip-flop is in state 1, the voltage at Y will be

high. Assume that all flip-flops are in state 0 (the transistors Q_0B, Q_1B, Q_2B and Q_3B are conducting) and that the sixteen successive input negative triggering pulses are applied at the T input of flip-flop B_0; then the waveforms shown in Figure 1-128 will appear at the outputs Y of the individual flip-flops. We may verify that the waveform chart of Figure 1-128 is correct by applying the following principles:

1. Flip-flop B_0 must make a transition at each externally applied pulse.

2. Each of the other flip-flops must make a transition when and only when the preceding flip-flop makes a transition from state 1 to state 0.

Thus the first external pulse applied to B_0 causes the flip-flop to make a transition from state 0 to state 1. As a result of this transition a positive voltage step is applied at the T input of B_1. Because of the arrangement of diodes in B_1 (see Figure 1-127), this positive step will not induce a transition in B_1. The overall result is that B_0 has changed its state to 1, and all other flip-flops remain in state 0, as in Figure 1-128. The second externally applied pulse causes flip-flop B_0 to return from state 1 to state 0. Flip-flop B_1 now receives a negative step voltage to which the flip-flop is sensitive, and it responds by making a transition from state 0 to state 1. Then the flip-flop B_2 receives a positive step from B_1, and hence does not respond to the transition in B_1. The overall result of the application of two input pulses is that flip-flop B_1 is in state 1 while all other flip-flops are in state 0. The remainder of the waveform chart can be verified in the similar manner.

We have used positive logic to describe our system as above. If the chain were constructed of flip-flops using *p-n-p* transistors, then the voltage level at Y corresponding to the 1 state would be lower than the voltage level corresponding to the 0 state. It would then be appropriate to describe the operation in terms of negative logic. The waveforms in

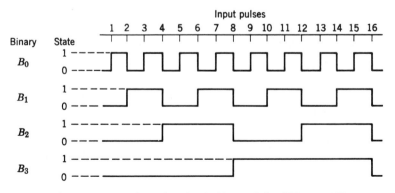

Fig. 1-128   Waveform chart for the binary chain of Figure 1-127.

Figure 1-128 would have to be inverted and the input waveform would have to be a positive pulse train. However, the principles given previously would continue to apply without modification.

Table 1-7
States of the Binaries

Number of Input Pulses	State of Binary			
	B_3	B_2	B_1	B_0
0	0	0	0	0
1	0	0	0	1
2	0	0	1	0
3	0	0	1	1
4	0	1	0	0
5	0	1	0	1
6	0	1	1	0
7	0	1	1	1
8	1	0	0	0
9	1	0	0	1
10	1	0	1	0
11	1	0	1	1
12	1	1	0	0
13	1	1	0	1
14	1	1	1	0
15	1	1	1	1
16	0	0	0	0
17	0	0	0	1

This table gives binary-to-decimal equivalences for the first 17 integers.

The states of the binaries listed in Table 1-7 may be verified directly by comparison with the waveform chart of Figure 1-128. For each binary one output pulse appears for two input pulses. A chain of n binaries will count up to the number 2^n before it resets itself into its original state. Such a chain is called a counter modulo 2^n. The number 2^n is called a dividing or scaling factor.

If we should differentiate each of the waveforms of Figure 1-128, a positive pulse would appear at each transition from 0 to 1 and a negative pulse at each transition from 1 to 0. If now we count only the negative pulses (the positive pulses may be eliminated by, say, using a diode), then it appears that each binary divides the number of negative pulses applied to it by 2. The four binaries together accomplish a division by a factor $2^4 = 16$. A single negative pulse will appear at the output for each 16 negative pulses applied at the input. A chain of n binaries used for this purpose of dividing or scaling down the number of pulses is known as a scaler. Thus a chain of four binaries constitutes a scale-of-16 scaler, etc. A scale-of-64 scaler followed by a mechanical register will be able to respond to the pulses of Geiger-Müller tubes which occur at a rate 64 times greater than the maximum rate at which the mechanical register will respond. The net count in such a case will be 64 times the reading of the mechanical register plus the count left in the scale-of-64 counter.

A counter which can be made to count in either the forward or reverse direction is called a reversible counter or a forward-backward counter. Forward counting is accomplished when the trigger input of a succeeding binary is coupled to the Y output of a preceding binary, as in Figure 1-127. The count will proceed in the reverse direction if the coupling is made instead to the $\overline{Y}$ output. For the reversing connection the following principles apply[1]:

1. Binary B_0 makes a transition at each externally applied pulse.
2. Each of the other binaries makes a transition when and only when the preceding binary goes from state 0 to state 1.

The Decade Counter with Feedback

The circuit of Figure 1-129 shows a scale-of-16 binary chain modified by feedback into a scale-of-10 binary chain. The signal at Y_3 of the last binary is differentiated by R and C. The negative pulse which results when this flip-flop goes from state 1 to 0 is ineffective because the coupling transistors Q_4 and Q_5 are virtually at cutoff. However, the posi-

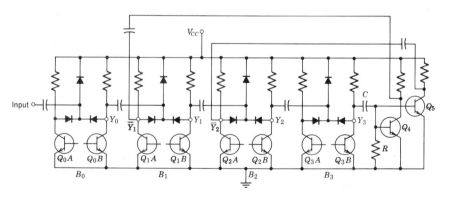

Fig. 1-129   A decade counter with feedback. The cross-coupling connections from collectors to bases have been omitted.

[1]If these principles are applied to any of the numbers in Table 1-7, the next smaller number in the table is the result.

tive pulses which result when B_3 goes from 0 to 1 are inverted and fed back to $\overline{Y}_1$ and $\overline{Y}_2$. Remember that a binary does not respond instantaneously to an input pulse and that there is some additional delay before the flip-flop output reaches full amplitude. This feature has been taken into account in the waveform chart of Figure 1-130 by drawing the binary transitions with a finite slope and by starting the transition in a succeeding binary only at the completion of the transition in a preceding flip-flop. The counting proceeds in normal fashion through the seventh input pulse. At the eighth pulse, flip-flop B_3 responds and negative pulses are fed back to $\overline{Y}_1$ and $\overline{Y}_2$. Since B_1 and B_2 are in state 0 after the eighth pulse, the right-hand transistors of these flip-flops are conducting. Therefore negative pulses at $\overline{Y}_1$ and $\overline{Y}_2$ (which are coupled to the opposite bases) will cause transitions in B_1 and B_2, respectively, and these flip-flops are forced back to state 1 by this feedback pulse. At the ninth pulse B_0 changes to state 1 and after the tenth pulse all flip-flops are again in state 0, so that the count is complete. Note that after the eighth pulse, before the feedback has had a chance to be effective, the chain of flip-flops is in the state $1000 = 8$ (see Figure 1-130). The feedback to flip-flop B_1 advances the count by $2^1 = 2$, which is the same as adding a 2 to the count. The feedback to B_2 advances the count by $2^2 = 4$, which is the same as adding another 4 to the count. The two feedback paths advance the count by $2 + 4 = 6$, which is the same as adding 6 to the count. Now the counter reads $1110 = 14$ (see Figure 1-130), and so two additional pulses are required to reset the counter to 0000. Thus the counter recycles at pulse 10 instead of 16.

Referring to the operating principle given above, we may use feedback in other ways to change the scale of a counter. However, feedback employed in such a decade counter may severely limit the maximum count-

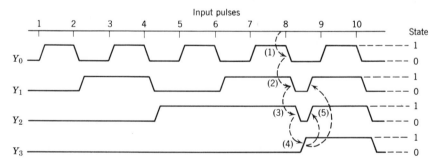

Fig. 1-130   Waveform chart for the feedback decade counter of Figure 1-129. (The dashed lines indicate the five delays 1 to 5.)

ing rate. For this reason some types of decade counters without feedback have been recently developed.

The Decimal Counter Without Feedback

Figure 1-131 shows an example of a decade counter that does not use feedback. The circuit consists of a chain of four flip-flops and a two-diode AND gate. Its counting sequence corresponds to the 8-4-2-1 code. Decimal 8 = 1000 binary, 4 = 100, 2 = 10, and 1 = 1 (see Table 1-7). This counter requires the following counting steps: 0000, 0001, 0010, 0011, 0100, 0101, 0110, 0111, 1000, 1001, and then return to 0000. Each of these four-digit groups indicates the states of the four binaries. The 8-4-2-1 code consists of straight binary counting up to 1001 = 9; the remaining states, 10 to 15, are merely omitted.

In the circuit of Figure 1-131 we use trigger steering by means of bias applied to the base trigger diodes. To describe the operation of this counter, let us use positive logic. With the *p-n-p* transistor polarity, this implies that logical 0 is represented by $\simeq -6$ volts, while logical 1 corresponds to ground potential. The transistors which are ON in the state 1 are identified by shading. The output points at their collectors are thus $\overline{Y}_0$, $\overline{Y}_1$, $\overline{Y}_2$, and $\overline{Y}_3$, while Y_0, Y_1, Y_2, and Y_3 appear at the opposite collectors.

The input positive pulses are applied to the first binary stage, Q_1 and Q_2, which scales in the normal manner. The system proceeds with carry pulses in the required straight binary counting sequence until state 1001 (= 9) is reached. Here the special decimal feature comes into play. The collector of Q_8 is now negative ($\overline{Y}_3 = 0$). The AND gate controlling point X (near the base of Q_4) thus draws X negative also, with the result that the base trigger to Q_4 is suppressed. The next input pulse changes the first flip-flop back to 0 ($Y_0 = 0$, $\overline{Y}_0 = 1$). The carry signal to the second flip-flop is blocked, however, because of the suppression just described. The carry pulse from the collector of Q_2 is instead applied directly to the base of Q_7, causing the last flip-flop to return to 0 ($Y_3 = 0$). We thus go from 1001 directly back to 0000.

Schmitt Trigger Circuit

A schmitt trigger circuit is shown in Figure 1-132. It is a regenerative bistable circuit whose state depends on the amplitude of the input voltage. The base of Q_1 is not involved in the regenerative switching. When the circuit switches between levels, the voltage on this base terminal does not change. The Schmitt circuit is therefore preferred for applications in which we desire to take advantage of this free terminal. A most important application of this circuit is its use as an amplitude comparator

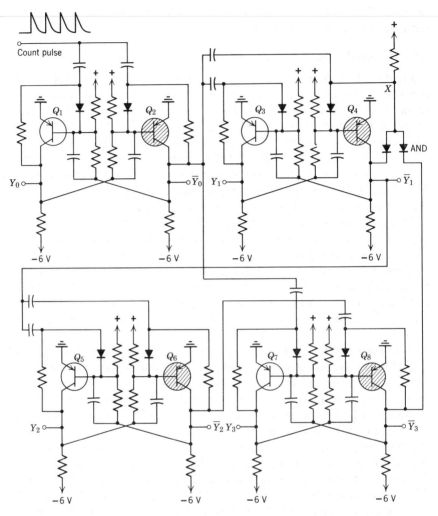

Fig. 1-131   Example of a decimal counter that does not use feedback.

to mark the instant when an arbitrary waveform attains some reference level. In a second application, the Schmitt circuit is used as a squaring circuit. This application is illustrated in Figure 1-133, where the arbitrary input signal has a large enough excursion to carry the input beyond the limits of the hysteresis[1] range $V_H \equiv V_1 - V_2$. The output is a square

---

[1]The difference between the input signal level required to trip a trigger circuit and the level which makes it recover.

wave, as shown, whose amplitude is independent of the amplitude of the input waveform.

In still another application the circuit is triggered between its two stable states by alternate positive and negative pulses. Thus if the input is biased to a voltage V between V_2 and V_1, and if a positive pulse whose amplitude exceeds $V_1 - V$ is coupled to the input, then Q_1 will conduct

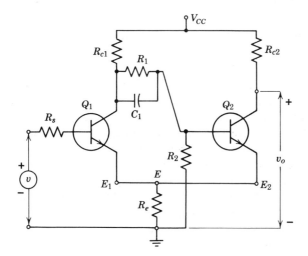

Fig. 1-132   A Schmitt trigger circuit (an emitter-coupled binary). The impedance of the input source is R_s.

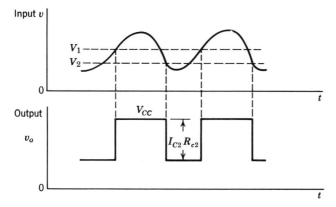

Fig. 1-133   A square-wave output from the Schmitt circuit in response to an arbitrary input signal.

and Q_2 will be driven to cutoff. If there now is applied a negative pulse whose amplitude is greater than $V - V_2$, the circuit will be triggered back to the state where Q_1 is OFF and Q_2 is conducting. This behavior is the same as that of a flip-flop. In the typical circuit of Figure 1-134, the frequency range is from 0 to 1 Mc; the output at collector of Q_2 has 2-V minimum level change; Q_1 always conducts if the input exceeds 6.8 V; Q_2 always conducts if the input is below 5.2 V. The ambient temperature is within 0 to 71°C.

1-20 MONOSTABLE MULTIVIBRATORS

The Basic Monostable Circuit

The one-shot or monostable circuit has only one permanently stable state and one quasi-stable (or unstable) state. On being triggered the circuit switches to its quasi-stable state where it remains for a predetermined time before returning to its original stable state. This makes the monostable multivibrator useful in standardizing pulses of random widths or in generating time-delayed pulses. The circuit is similar to that of a flip-flop, except that one cross-coupling network permits ac coupling only. Therefore, the flip-flop can only remain in its quasi-stable state until the circuit reactive components discharge.

Figure 1-135 shows a basic monostable multivibrator using two *n-p-n* transistors, Q_1 and Q_2. The bias network holds Q_2 in saturation and Q_1

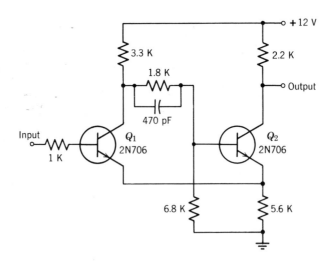

Fig. 1-134   A typical Schmitt trigger circuit. (From *Transistor Manual*, 7th ed., General Electric Company, 1964.)

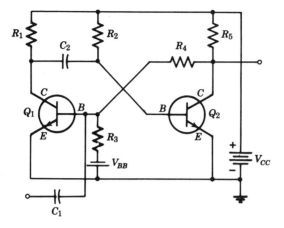

Fig. 1-135   A collector-coupled monostable multivibrator.

at cutoff during the quiescent (steady-state) period. When a positive pulse is applied through the coupling capacitor C_1, however, Q_1 begins to conduct. The decreasing collector voltage of Q_1 (coupled to the base of Q_2 through C_2) causes the base current and collector current of Q_2 to decrease. The increasing collector voltage of Q_2 (coupled to the base of Q_1 through resistor R_4) then increases the forward base current of Q_1. This regeneration rapidly drives Q_1 into saturation and Q_2 into cutoff. The base of Q_2 at this point is at a negative potential almost equal to the magnitude of the supply voltage V_{CC}. Capacitor C_2 then discharges through resistor R_2 and the low-saturation resistance of Q_1. As the base potential of Q_2 becomes slightly positive, transistor Q_2 again conducts. The decreasing collector voltage of Q_2 is coupled to the base of Q_1 and Q_1 is driven into cutoff while Q_2 becomes saturated. This stable condition is maintained until another pulse triggers the circuit. The duration of the output pulse is primarily determined by the time constant R_2C_2 during discharge.

The Emitter-coupled Monostable Multivibrator

The emitter-coupled monostable multivibrator is shown in Figure 1-136. In the stable state Q_1 is cut off and Q_2 is in saturation. The bias voltage V is obtained from the divider R_1R_2 across the supply voltage V_{CC}. The feedback is provided through a common emitter resistor R_e. The signal at the collector of Q_2 is not directly involved in the regenerative loop. Therefore this collector makes an ideal point from which to obtain an output voltage waveform. The base of Q_1 is a good point at which to in-

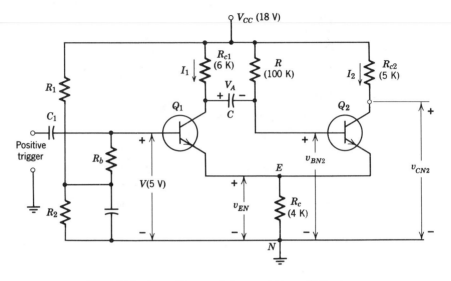

Fig. 1-136   An emitter-coupled monostable multivibrator.

ject a triggering signal, since this electrode is coupled to no other in the circuit. Thus the trigger source cannot load the circuit. When Q_2 goes OFF and Q_1 goes ON, Q_1 operates with a substantial emitter resistance. This emitter resistance will serve to stabilize the collector current I_1. Hence the time T of the quasi-stable state may be controlled through I_1. The current I_1 can be adjusted through the bias voltage V, and it turns out that T varies rather linearly with V. Thus the emitter-coupled mono-stable circuit makes an excellent gate waveform generator, whose width is easily and linearly controllable by means of an electrical signal voltage. The waveforms of the emitter-coupled monostable circuit (Figure 1-136) are shown in Figure 1-137. Some modified emitter-coupled monostable multivibrators have been widely used in the Geiger-Müller tube survey meters.

1-21 BINARY ADDERS

Binary Addition

In a digital computer the basic arithmetic operations are addition and subtraction; multiplication is, essentially, repeated addition and division is, essentially, repeated subtraction. In making binary addition, we must add not only the digit of like significance of the two numbers to be summed but also the carry bit (should one be present) of the next lower significant

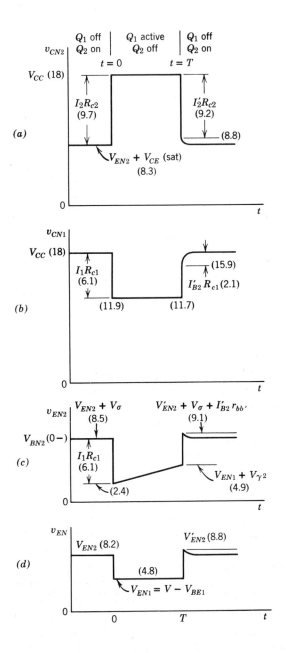

Fig. 1-137 Waveforms of the emitter-coupled monostable circuit shown in Figure 1-136.

147

digit. This operation may be carried out in two steps. First the two bits corresponding to the 2^k digit are added together, and then the resultant is added to the carry from the 2^{k-1} bit.

A Half-Adder-Subtractor

A two-input adder is known as a half-adder, since it requires two such half-adders to complete an addition. A half-adder-subtractor has two inputs A and B, representing the bits to be added, and three outputs $-D$ (for the digit of the same significance as A and B represent), C (for the carry bit), and P (for the borrow bit). In a half-adder D and C are used, while in a half-subtractor D and P are used.

Figure 1-138 shows a half-adder-subtractor with its symbol and truth table. The D column gives the sum of A and B as long as the sum can be represented by a single digit. When the sum is larger than this, D gives the digit in the result, which is of the same significance as the individual digits being added. Thus, in the first three rows of the truth table, D gives the sum of A and B directly. Since the decimal equation "1 plus 1 equals 2" is written in the binary form as "01 plus 01 equals 10" (see Table 1-6), then in the last row $D = 0$. Because a 1 must now be carried to the place of next higher significance, $C = 1$. Finally, where subtraction of B from A is contemplated, the P (borrow) column gives the digit which must be borrowed from the place of next higher significance when B is larger than A, as in the second row of Figure 1-138c.

The Parallel Binary Adder

Two multidigit numbers may be added in parallel (all columns simultaneously) by using a parallel binary adder. For an n-digit binary number there are (in addition to a common ground) n signal leads in the computer for each number. The kth line for number A (or B) is excited by A_k (or B_k), the bit for the 2^k digit ($k = 0, 1, 2, \ldots, n$). Figure 1-139 shows a parallel binary adder. Each digit except the least-significant one (2^0) requires a full adder which consists of two half-adders in cascade. The sum digit for the 2^0 bit is $S_0 = D_0$ of a half-adder because there is no carry to be added to A_0 plus B_0. The sum S_k ($k \neq 0$) of A_k plus B_k is made in two steps. First the digit D_k is obtained from one half-adder, and then D_k is summed with the carry C_{k-1}, which may have resulted from the next lower place. As an example, consider $k = 2$ in Figure 1-139. There the carry bit C_1 may be the result of the direct sum of A_1 plus B_1 if each of these is 1. This first carry is called C_{11} in Figure 1-139. A second possibility is that $A_1 = 1$ and $B_1 = 0$ (or vice versa), so that $D_1 = 1$, but that there is a carry C_0 from the next lower significant bit. The sum of $D_1 = 1$ and $C_0 = 1$ gives rise to the carry bit designated C_{12}. It should be clear that C_{11} and C_{12} cannot both be 1, although they will both be 0 if

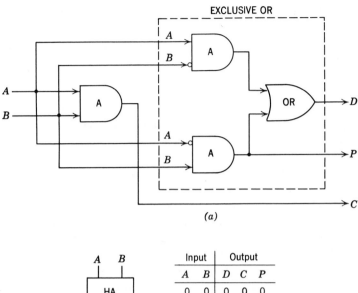

(a)

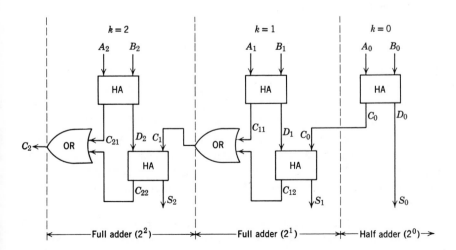

Fig. 1-139   A parallel binary adder consisting of half-adders.

$A_1 = 0$ and $B_1 = 0$. Since either C_{11} or C_{12} must be transmitted to the next stage, an OR gate must be interposed between stages, as shown in Figure 1-139. This circuit is equally effective for subtraction, provided that the borrow bit P is used in place of the carry C.

The Serial Binary Adder

Two multidigit numbers may be added serially (one column at a time) by using a serial binary adder. In a serial adder the inputs A and B are synchronous pulse trains on two lines in the computer. Figures 1-140a and b show typical pulse trains representing, respectively, the decimal numbers 13 and 11. Pulse trains representing the sum (24) and difference (2) are shown in Figures 1-140c and d, respectively. A serial adder is a device which will take as inputs the two waveforms of Figures 1-140a and b and deliver the output waveform in Figure 1-140c. Similarly, a subtractor will yield the output shown in Figure 1-140d.

It should be noted that at any instant of time we must add (in binary form) to the pulses A and B (Figures 1-140a and b) the carry pulse (if any) which comes from the resultant formed one period T earlier. The carry pulse may arise from the direct sum of two digits (each 1) or from the addition of digits 1 and 0 and a carry 1 from the preceding interval. The logic outlined above is performed by the full-adder circuit of Figure 1-141, which consists essentially of two half-adders in cascade. The time delay TD of the electromagnetic delay line is equal to the time T between pulses. Therefore, the carry pulse (from either of the two sources mentioned above) is delayed a time T and added to the digit pulses in A and B, which is exactly as it should be. Serial addition is much cheaper but slower than parallel operation.

PROBLEMS

1-1 A 10-Hz (cycles per second) symmetrical square wave whose peak-to-peak amplitude is 4 V is impressed on a high-pass RC network whose lower 3-dB frequency is 5 Hz. Calculate and sketch the output waveform. In particular, what is the peak-to-peak output amplitude?

1-2 A square wave whose peak-to-peak value is 2 V extends ± 1 V with respect to ground. The half period is 0.1 sec. This voltage is impressed on an RC differentiating circuit whose time constant is 0.2 sec. What are the steady-state maximum and minimum values of the output voltage?

1-3 A symmetrical square wave whose average value is zero has a peak-to-peak amplitude of 10 V and a period of 2 μsec. This waveform is applied to a low-pass RC network whose upper 3-dB frequency is

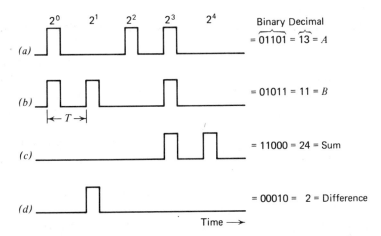

Fig. 1-140 Pulse waveforms (*a*, *b*, *c*, *d*) representing number *A*, number *B*, sum, and difference, respectively.

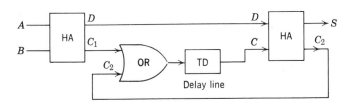

Fig. 1-141 A serial binary adder consisting of two half-adders in cascade.

$1/2\pi$ MHz. Calculate and sketch the steady-state output waveform. In particular, what is the peak-to-peak output amplitude?

1-4 A symmetrical square wave whose peak-to-peak amplitude is 20 V and whose average value is zero is applied to an *RC* integrating circuit. The time constant equals the half-period of the square wave. Find the peak-to-peak value of the output amplitude.

1-5 A transformer has the following parameters: $L = 6.0$ mH, $\sigma = 30$ μH, $C = 100$ pF, $R_1 = R_2 = 500$ Ω, and $n = 1$. Find and plot the response to a 1-μsec 20-V pulse.

1-6 A type RG-59/U coaxial cable has a capacitance of 20 pF/ft and a characteristic impedance of 73 Ω. Find the length required for a 0.4-μsec delay.

1-7 A coaxial cable with a nylon dielectric ($\epsilon_r = 3.00$) has a characteristic impedance of 200 Ω.

    (a) How long a cable is needed to give a one-way delay of 40 nsec?

    (b) What is the ratio of the outer to the inner radius?

1-8 (a) A transistor operates in the common-emitter configuration and drives a 2-kΩ load. The h parameters (at $I_E = 1$ mA) are $h_i = 1.4$ KΩ, $h_r = 3.4 \times 10^{-4}$, $h_f = 44$, $1/h_0 = 37$ KΩ. Calculate the current gain A_I, the input impedance Z_i, and the voltage gain A_V.

    (b) If the transistor is reconnected to operate as an emitter follower with the same 2-kΩ resistor in the emitter circuit, calculate A_I, Z_i, and A_V.

1-9 A silicon transistor operating at room temperature and at an emitter current of 2.0 mA has hybrid-π parameters $r_{bb'} = 50$ ohms, $r_{b'e} = 2$ K, $r_{ce} = 100$ K, and $r_{b'c} = 2$ MΩ. Find the common-emitter hybrid parameters.

1-10 Consider the emitter-coupled monostable multivibrator whose components and supply voltages are indicated in Figure 1-136. Assume that $V = 5.0$ V and that each of the Ge transistors has $h_{FE} = 50$ and $r_{bb'} = 200$ Ω. Calculate the voltage levels of the waveforms in Figure 1-137.

    Hint: (a) Draw an equivalent circuit of Q_2 (in saturation) in the stable state ($t < 0$). Use Table 1-5. $I_2 = 1.95$ mA; $I_{B2} = 0.095$ mA; min current for Q_2 in saturation $= I_2/h_{FE} = 0.039$ mA; $V_A = 9.5$ V. (b) Draw an equivalent circuit of Q_1 during the quasi-stable state when Q_2 is OFF. Calculate at $t = 0+$. $I_{C1} = 1.18$ mA; $I_R = 0.156$ mA; $I_1 = 1.02$ mA.

    Ans.: For t < 0, $v_{CN1} = V_{CC} = 18$ V; $v_{CN2} = 8.3$ V; $v_{EN} \equiv V_{EN2} = 8.2$ V; $v_{BN2} = 8.5$ V.

    At $t = 0+$, $v_{CN1} = 11.9$ V; $v_{CN2} = 18$ V; $v_{EN} = V_{EN1} = 4.8$ V; $v_{BN2} = 2.4$ V.

REFERENCES

Millman, J., and H. Taub, *Pulse, Digital, and Switching Waveforms,* McGraw-Hill, New York, 1965.

Littauer, R., *Pulse Electronics,* McGraw-Hill, New York, 1965.

Shea, R. F., *Transistor Circuit Engineering,* Wiley, New York, 1958.

Chase, R. L., *Nuclear Pulse Spectrometry,* McGraw-Hill, New York, 1961.

Greiner, R. A., *Semiconductor Devices and Applications,* McGraw-Hill, New York, 1964.

Price, W. J., *Nuclear Radiation Detection,* 2d ed., McGraw-Hill, New York, 1964.

Yanof, H. M., *Biomedical Electronics,* Davis, Philadelphia, 1965.

Transistor Manual, 7th ed., General Electric Company, 1964.

Transistor Manual, SC-13, Radio Corporation of America, 1967.

Instruction Manual, *Operating and Maintenance Instructions for* 400 *Channel Analyzer System, Model* 20631/2, Radiation Counter Laboratories, Inc.

Fairstein, E., and J. Hahn, *Nuclear Pulse Amplifiers*—Fundamentals and Design Practice, Parts I, II, III, *Nucleonics,* **23**, 7, 56 (July, 1965); 9, 81 (Sept., 1965); 11, 50 (Nov., 1965).

Proposed Standard Definitions of General (Fundamental and Derived) Electrical and Electronics Terms, *IEEE* No. 270, Institute of Electrical and Electronics Engineers, New York, 1966.

Malmstadt, H. V., and C. G. Enke, *Electronics for Scientists,* Benjamin, New York, 1963.

Millman, J., and C. C. Halkias, Electronic Devices and Circuits, McGraw-Hill, New York, 1967.

CHAPTER **2**

PULSE AMPLIFIERS AND SINGLE-CHANNEL ANALYZERS

2-1 OPERATION OF CURRENT-MODE AND VOLTAGE-MODE SIGNALS

The Input Signal Dependent of Detector Characteristics

The basic function of nuclear pulse amplifiers is to take the low-level pulse issuing from the radiation detector and transform it, through amplification and pulse shaping, into an output pulse more suitable for measurement and analysis. In order for the information contained in the original pulse to be preserved, the amplifier must not only retain a linear relationship between the output pulse height and the input pulse but must also be capable of providing extremely accurate timing information. The shape, magnitude, and noise content of the input signal to the main amplifier are determined by the detector and preamplifier. The electrical signal in any type of detector is a current or an impulse of charge. The total charge generated by each photon or particle absorbed in the detector is directly proportional to the absorbed energy (or energy loss). The signals produced by various detectors fall in the range between 10^{-15} and

10^{-10} coul/pulse. The duration of the current varies with the detector and with the nature and energy of the radiation. The shortest durations occur in plastic scintillators and narrow-depletion-depth semiconductor detectors (0.1 to 10 nsec), and the longest occur in the gas-filled detectors (0.01 to 5 μsec). NaI(Tl) and CsI(Tl) scintillators are intermediate with total current durations of 0.75 to 1.5 μsec.

In semiconductor radiation detectors, the energy-conversion factors ϵ are 2.9 and 3.6 eV per electron-hole pair for germanium and silicon, respectively. No charge multiplication occurs within the detector. In scintillation detectors, the effective ϵ is 3 to 30 keV per electron, with multiplication factors as high as 10^7 occurring in the photomultiplier. In the gas-filled detectors, the conversion factor is 25 to 35 eV per ion pair. No multiplication occurs in ion chambers, but multiplication factors as great as 10^4 are common for proportional counters.

There are two areas of signal processing in which the distinction between current- and voltage-mode operation can be made. These will be described in the following section.

Current- and Voltage-Mode Operation

It should be noted that the instantaneous amplitude of the current is not proportional to the energy loss in the detector; it is the charge, or integral, of the total current which bears the desired proportionality. Thus at the detector or in the amplifier, an integration must be performed to obtain accurate energy information.

If the detector time constant is very small in comparison with the duration of the current, the shape of the current waveform is preserved and a current-mode signal results. On the other hand, if the time constant is very large, the current waveform is integrated and a voltage-mode signal is obtained. The height of the resulting voltage step is Q/C, where Q is the charge (in coulombs) released in the detector and C is the capacitance (in farads) which shunts it. The step voltage decays exponentially with the detector time constant. The rise time is nearly equal to the duration of the current and is virtually independent of the detector time constant.

Notice that the terms "current- and voltage-mode signals" are completely unrelated to the terms "current amplifier" and "voltage amplifier." A current amplifier is one in which the input impedance is low and the output impedance is high. For a given input signal power, the low impedance results in a high current and low voltage. The converse is true of voltage amplifiers. The current-mode or voltage-mode signal may be used with either a current amplifier or a voltage amplifier.

Three Distinct Amplifier Systems

Pulse amplifier systems are of three types. The first is the voltage-mode system. Here the signal is integrated at the detector to produce a tail pulse.[1] This pulse is then differentiated late in the preamplifier or early in the main amplifier and, possibly, a second time later in the main amplifier. Perhaps the principal disadvantage of this voltage-mode system is that at high counting rates pileup can become a problem, even in the early stages of amplification. Pileup means the overlapping of a sequence of pulses to produce a staircase effect, and it is unavoidable with random spacing between successive pulses. When this problem occurs, the second type of signal processing, that is, current-mode, may be advantageous.

In this system a low-valued detector load resistor, real or virtual, is used to preserve the shape of the current signal. The detector is followed by a preamplifier and main amplifier having rise times comparable to the duration of the current. At or near the output stage, a multiple integration is introduced. The first integration converts the current-mode signal to a voltage-mode tail pulse, and the second and successive integrations round the leading edge and introduce symmetry to the pulse shape. In this system, before the signal is converted to the voltage mode, logical operations can be performed on the current mode signal with a minimum of interference between adjacent pulses. This second system, however, is usable only when the energy resolution is not determined by the signal-to-noise ratio of the preamplifier.

When both the pileup and signal-to-noise ratio in the early stages are critical to the performance, a third system may be used to advantage. It consists of a voltage-mode preamplifier followed by a current-mode main amplifier. In this third system the detector current is first converted to a voltage-mode tail pulse through the use of a high-valued detector load resistor and then converted back to a current pulse by a heavy differentiation at the output of the preamplifier.

Signal Widths in Current-Mode Amplifiers

In the current-mode amplifier, the minimum current-mode signal duration will be determined by the longest of the following three parameters: (a) duration of the detector current, (b) amplifier rise time, and (c) shortest signal decay time in the system. Which of these should control depends upon the detector current duration and the use to which the signal will be put. If the purpose is to study the detector waveforms, it is mandatory that the amplifier rise time and detector time constant be less

---

[1]A tail pulse is a pulse having a short rise time and a relatively long exponential decay.

than the detector current duration. (This condition will be impossible to fulfill if the duration is less than about 2 nsec.) If logical operations must be performed on the signal before it is integrated, the signal duration should be compatible with the speed of the logic circuits. The signal decay time constant should be approximately equal to the amplifier rise time, so that an optimum situation will exist. Any signal having a duration shorter than the pulse which leaves the output terminal of the main amplifier will be considered a current-mode signal.

Current and Voltage Types of Amplification

For a given parasitic capacitance of the amplifier circuit, the time required to charge it is directly proportional to the voltage swing and inversely proportional to the signal current. Also, the amplifier speed is inversely proportional to the charge time. Thus the low-voltage high-current operation of current amplifiers results in a distinct speed advantage. This operation makes it possible to use the high-frequency (fast) transistors with low breakdown voltage ratings. By the selection and arrangement of the feedback and load resistors, the same basic configuration can serve as a current amplifier, voltage amplifier, or converter between the two.

In the amplifier stage of Figure 2-1, the voltage amplification between point X and terminal 2 is $-A$. The negative feedback network consists of R_1 and R_2. For a given output voltage at terminal 2, the signal at X is $-1/A$ times as great. If the magnitude of A is much greater than 1, the signal at X is much less than that at the input or output terminals. If the signal source at terminal 1 is a current generator, the voltage developed there is $v_i = i_i R_1$, where i_i is the signal current and R_1 is the input resistor. On the other hand, if the source is a voltage generator, the current which flows is $i_i = v_i/R_1$. For current amplification, R_1 is usually chosen to match the connecting cable impedance within the range of 50 to 125 ohms. For voltage amplification, R_1 is usually 500 to 2000 ohms. The signal at terminal 2 is a voltage of the magnitude given by

$$v_0 = v_i \frac{R_2}{R_1}$$

or

$$v_0 = \frac{v_i}{R_1} R_2 = i_i R_2. \tag{2-1}$$

Equation (2-1) indicates that the amplifier of Figure 2-1 can be used as a current-to-voltage converter.

The emitter current of Q_1 is $i_e = v_0/R_2 \parallel R_3$, where v_0 is the output voltage at terminal 2 and $R_2 \parallel R_3$ is the effective resistance between terminal

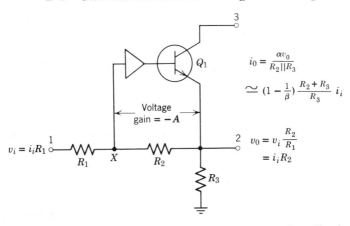

$$i_0 = \frac{\alpha v_0}{R_2 \| R_3}$$

$$\simeq (1 - \frac{1}{\beta}) \frac{R_2 + R_3}{R_3} i_i$$

$$v_0 = v_i \frac{R_2}{R_1}$$

$$= i_i R_2$$

$$v_i = i_i R_1$$

Fig. 2-1   Amplifier stage illustrating both current and voltage types of amplification as well as current-to-voltage conversion.

2 and ground (or virtual ground). The signal at terminal 3 is a current, and its magnitude is given by

$$i_0 = \alpha \, i_e = \frac{\alpha v_0}{R_2 \| R_3}.$$

Substituting $\alpha = \beta/(1 + \beta) \simeq 1 - 1/\beta$, $v_0 = i_i R_2$, $R_2 \| R_3 = R_2 R_3/(R_2 + R_3)$, we obtain the current gain

$$A_I = \frac{i_0}{i_i} \simeq \left(1 - \frac{1}{\beta}\right) \frac{R_2 + R_3}{R_3}, \tag{2-2}$$

where the $(1 - 1/\beta)$ term accounts for the signal loss in the base of Q_1. The transistor current amplification factor β is usually between 25 and 200. The main disadvantage to using terminal 3 as a current output terminal is that β[1] is temperature dependent and cannot be stabilized by feedback. However, this temperature effect is eliminated by using terminal 2 as output terminal and by making R_3 the input resistor of a following stage of similar design. In this case, the output voltage (v_0) of the first stage becomes the input voltage to the second; then $v_0 = i_i R_2$, $i_0' = v_0/R_3$, and we have the current gain

$$\frac{i'_0}{i_i} = \frac{R_2}{R_3}, \tag{2-3}$$

where i'_0 is the input current to the second stage.

The voltage-mode signal processing is more popular than the current-

[1]Typical values for the temperature coefficient of β vary from 0.5 to 2 percent per °C. In the worst case, $\beta = 25$ and $\Delta\beta/\beta = 2$ percent per °C, the change in $(1 - 1/\beta)$ is 0.08 percent per °C.

mode. The voltage pulses can be distorted in height by nonlinearity, pulse-shaping, pileup, overload, and noise.

2-2 VARIOUS SOURCES OF PULSE-HEIGHT DISTORTION

Nonlinearity and Control of Nonlinearity

In that part of the amplifier following the pulse-shaping networks, integral nonlinearity affects the energy axis calibration, and differential nonlinearity affects the shape of individual peaks. Numerically, the differential nonlinearity is always the greater of the two.

Nonlinearity occurs in tubes and transistors subjected to a current swing. A transistor, like a tube, has a transconductance (g_m) directly proportional to the current through it. In a high-level amplifier stage the current swing may be quite large. With a low-impedance driving source, the voltage gain of a single transistor stage (common emitter) is approximately $g_m R_L$, where R_L is the load resistance. With a relative high-impedance driving source, the voltage gain is approximately $\beta R_L / R_S$ [see (1-99)], where R_S is the driving source resistance. The value of g_m can be assumed proportional to the quiescent operating current [(1-100)]. β is also a function of operating current; however, the variation of β with current is considerably less than that of g_m. In most actual circuits the driving source impedance is somewhere between the two limits defined above. As a result, in assuming the voltage gain to be proportional to g_m we are assuming a worst-case dependence of gain upon operating current.

A frequently used technique for reducing the current variation is "boot-strapping." A boot-strapped circuit consists of an amplifier and an emitter follower (or cathode follower), as in Figure 2-2a. The output of the follower (Q_2) is fed back to the load resistor (R_2) of the amplifier (Q_1) for the purpose of keeping the voltage drop across R_2 (between X and Y) nearly constant, so that the current and gain of Q_1 remain nearly unchanged. The large current variation through the upper portion of R_2 is absorbed by Q_2, which, through the feedback inherent in an emitter follower, does not allow much of a gain shift. However, at high count rates (especially under overload conditions), a duty-cycle (see page 166) shift will occur at point Y, so that the gain shift in Q_1 will be unavoidable.

This duty-cycle effect is eliminated by using the circuit shown in Figure 2-2b. In this circuit Q_3 acts as a constant-current load for Q_1. The current is set by the resistor R_a and the base voltage at Q_3. Therefore the collector of Q_1 can swing over nearly the full range of power supply voltage with only a negligibly small change in collector current.

In practice another emitter follower must be added between the output

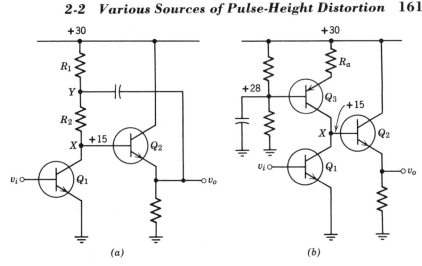

Fig. 2-2 Arrangements for reducing the current variation: (*a*) bootstrapped amplifier; (*b*) amplifier with constant current load. In practice an additional emitter follower is connected to the output terminal shown in the figures.

terminal shown in Figure 2-2*a* or *b* and the point at which the external load is applied. Otherwise loading at the output terminal shown in the figures will be reflected in loading at point *X*.

Influence of Pulse Shaping on Pulse Height

Assume that the output signal from the detector is integrated to produce a tail pulse before it feeds into the amplifier system. The rise time is determined by the detector characteristics, and the tail by the resistance and capacitance appearing at the terminals of the detector. For the best possible signal-to-noise ratio, the capacitance must be kept as low as possible and the resistance as high as possible. The resulting tail is undesirably long, but it may be shortened with a strategically located differentiating network. The time constant of the differentiator influences the signal-to-noise ratio, the count-rate limitations of the amplifier, and, in conjunction with other aspects of the shaping networks, the linearity of pulse-height analyzer response. When a voltage step or a tail pulse is applied to the input of a differentiator, the resulting output signal is proportional to the height of the step but of much shorter duration. The resulting width is the resolving time T_W of the differentiator. T_W equals the ratio of the pulse area to height.

If the amplifier has an *RC* interstage coupling network in addition to the differentiator, there will be no dc component in the output signal; hence the primary pulse (first part of a bipolar pulse) must be followed by an undershoot having the same area, as shown in Figure 2-3. The

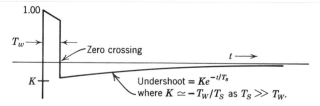

Fig. 2-3   Pulse with undershoot resulting from use of idealized differentiator and single *RC* interstage coupling. The undershoot has the same area as that of the primary pulse.

undershoot is exponential, with a decay constant determined by the secondary time constant (T_S), and with a peak amplitude relative to that of the primary pulse which asymptotically approaches T_W/T_S for $T_S >> T_W$ (see Figure 2-3). The secondary time constant T_S is the second shortest time constant in the amplifier. Note that the total pulse shown has a single zero crossing or crossover point. The pulse applied to the differentiator is the output signal of a detector which has the same waveform as that obtained from a perfect step-function generator followed by an interstage coupling network.

If there are two *RC* coupling networks in addition to the differentiator, an overshoot, having an amplitude approximately equal to $(T_W/T_S)^2$, will follow the undershoot, where the time constant of the second network may be any value between T_W and T_S, if the condition $T_S >> T_W$ still obtains. The total pulse will have two zero crossings.

If in a system of N coupling networks two of them have the same time constant and this time constant is appreciably shorter than any of the others, the detector pulse will be doubly differentiated and exhibit a large undershoot followed by a much smaller (or negligible) overshoot, as in Figure 2-4.

Pileup Effects

ONE OF THE CAUSES OF SPECTRAL-LINE BROADENING.   Pileup occurs at all signal points in the preamplifier and amplifier, and is one of the causes of spectral-line broadening. The mechanism by which broadening occurs in the preamplifier (before the differentiator) is different from that in the main amplifier (after the differentiator). This difference is explained with the help of the distinct waveforms in Figure 2-5.

PREAMPLIFIER PILEUP.   In the preamplifier, pileup may drive the output stage over a wide dynamic range. Nonlinearity causes the steps at the extremes of the range to have different heights, thereby broadening an otherwise monoenergetic pulse-height spectrum.

Figure 2-6*a* shows the input stage of a typical charge-sensitive pre-

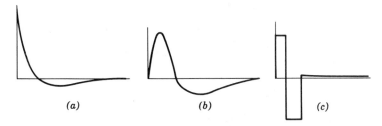

Fig. 2-4 Doubly differentiated pulse shapes: (a) double *RC*; (b) double *RC* with single *RC* integrator; (c) double delay line. The undershoot can vary from a minimum of 13.5 percent for cascaded *RC* differentiators (a) to a maximum of 100 percent for delay lines (c), through the intermediate value of 34.5 percent for cascaded *RC* networks with a single *RC* integrator (b).

amplifier. For an input signal which is an impulse of charge Q, the output pulse is given by

$$v_o(t) = \frac{Q}{C}\, \epsilon^{-t/RC},\qquad (2\text{-}4)$$

where Q is the input charge in coulombs, C is in farads, and R is in ohms. At a count rate of 1000 cps, few pulses will fail to overlap each other. The root-mean-square value of the output pulse is the root-mean-square fluctuation of the instantaneous base line, and it is given by

$$v_{0(\text{rms})} = \frac{Q}{C}\left(\frac{nRC}{2}\right)^{1/2},\qquad (2\text{-}5)$$

where n is the count rate. Equation (2-5) is derived from Campbell's theorem of the mean square:

$$v_{0(\text{rms})} = \left[\, n \int_0^\infty |f(t)|^2\, dt\,\right]^{1/2},\qquad (2\text{-}6)$$

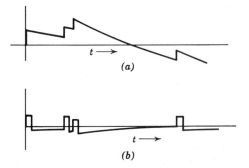

Fig. 2-5 Waveforms (a) in preamplifier and (b) in main amplifier.

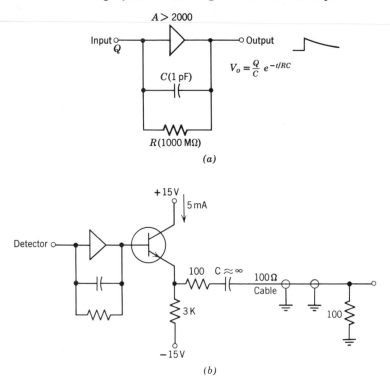

Fig. 2-6 (a) Input stage of a typical very-low-noise charge-sensitive preamplifier; (b) pre-amplifier circuit with emitter-follower output stage (see Prob. 2-2).

where $f(t)$, in this case, corresponds to $v_0(t)$ of (2-4). Approximately 68 percent of the pulses will be subjected to a fluctuation not exceeding the value of $v_{0(rms)}$ given by (2-5); the remaining 32 percent will suffer a greater base-line fluctuation (see Prob. 2-2). The degree of line broadening depends on the shape and magnitude of the preamplifier linearity curve over this voltage swing.

Several models of commercial preamplifiers have a second stage with a gain of 2 to 16 coupled to the first through a time constant which is usually about 50 μsec. With two cascaded time constants, the root-mean-square fluctuation of the base line due to pileup in the second stage is given by

$$v_{0(rms)} \simeq A \frac{Q}{C} \left[\frac{nT_1 T_2}{2(T_1 + T_2)} \right]^{1/2} \qquad (2-7)$$

where A is the second-stage gain, Q is the charge per pulse released by the detector, C is the feedback capacitor in the charge-sensitive stage, n is the count rate in cps, T_1 is the time constant of the feedback network in the charge-sensitive stage, and T_2 is the coupling time constant between first and second stages. If $T_1 > 10T_2$, (2-7) simplifies to

$$v_{0(\text{rms})} \simeq A \frac{Q}{C} \left[\frac{nT_2}{2} \right]^{1/2} \tag{2-8}$$

with an error of less than 1.0 percent.

MAIN AMPLIFIER PILEUP. In the main amplifier, after at least a single differentiation, pileup occurs on the undershoots, and, to a lesser degree, on the primary pulses. Since most pulse-height analyzers measure pulse heights from the ground level, those pulses which fall on the undershoots of earlier ones will appear to have lower amplitudes; the statistical fluctuations in the magnitude of the base-line depression result in dispersion of the measured pulse heights. If the system is singly differentiated and no base-line restoration is used in the pulse-height analyzer, spectral dispersion and peak shift will depend on the mean spacing between pulses relative to the time constant of the undershoot. This mean spacing corresponds to the reciprocal of the mean count rate. When the mean spacing is large compared with the secondary time constant T_S, a spectrum of pulse heights will show many of the pulses to be unaffected by the presence of others. If the mean spacing between pulses is not so large, then the succeeding pulse will fall on the undershoot of the preceding one. These cases are explained below.

With single differentiation, the peak shift due to pileup is shown in Figure 2-7, where the mean spacing between pulses b and c is assumed to be large compared with the secondary time constant. Pulse b falls on the undershoot of pulse a and thus appears to have lower amplitude than pulse a. Pulse b has decayed to the level of the base line and so has no influence on pulse c. The spectra of pulse heights corresponding to pulses b and c of Figure 2-7 are shown in Figure 2-8. Here the upper pulse height (energy) limit of peak c is sharply defined and corresponds to those pulses which start from the nondisplaced base line, while the lower edge of the spectrum is smeared, and corresponds to those pulses which fall at various points along the tails of earlier ones. In this case the peak is undisplaced, but the centroid is shifted toward lower energy by a fractional amount numerically equal to the duty cycle (nT_W). At high count

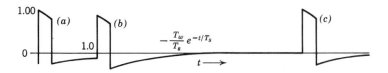

Fig. 2-7 Peak shift due to pileup dependent on the mean spacing between pulses.

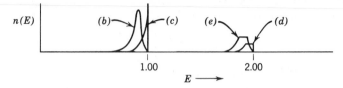

Fig. 2-8   Spectra of pulse heights corresponding to pulse shapes of Figure 2-7. E (energy) represents pulse heights; $n(E)$ represents count rates. Count rates contained in spectra of (b) and (e) are much higher than in (c) and (d).

rates, all but a vanishing fraction of the pulses are affected by earlier ones, so that both the peak and the centroid of the distribution are shifted, as can be seen from peak b in Figure 2-8.

The rms value of the spectral line dispersion due to the undershoot can be calculated from Campbell's theorem [see (2-6)], where $f(t) = (T_W/T_S)\,\epsilon^{-t/T_S}$. This rms value relative to the undistorted pulse height is then given by

$$\text{Relative dispersion, rms} = 0.707\,[(T_W/T_S)\,nT_W]^{1/2}, \qquad (2\text{-}9)$$

where T_W/T_S is the fractional undershoot and nT_W is the duty cycle.[1] For example, assume a resolving time of 1 μsec, a secondary time constant of 50 μsec, and a count rate of 10,000 cps. The undershoot will be $T_W/T_S = 1/50 = 2$ percent, the duty cycle will be $nT_W = 10^4 \times 10^{-6} = 1$ percent, and the rms dispersion will be 1 percent. Thus the centroid of the spectrum will be shifted lower in energy by 1 percent. Since we assume that the relative height of the undershoot $= T_W/T_S$, (2-9) is valid (within 5 percent) only for the condition $T_S \geq 10T_W$.

The spectra at points d and e in Figure 2-8 correspond to the sum spectra at low and high count rates, respectively, that is, to the spectra of those pulses that overlap, not on the tails, but directly in the region of the primary pulse. The area under the sum peaks is equal to the probability of primary pulse coincidence, and this probability is equal to the duty cycle. The shapes of the sum spectra are strongly dependent on the shape of the primary pulses.

We have discussed the pileup effect with single differentiation as above. Now, let us see how about the overlapping with double differentiation. In this case the duration of the undershoot is greatly reduced and the sum spectrum becomes the dominant form of dispersion. It is possible to work the system at much higher count rates than would be possible with single differentiation. To estimate the shape of the sum spectrum a quantitative method, which is given below, is used.

[1] The duty cycle is also defined as the ratio of resolving time to the mean spacing between pulses.

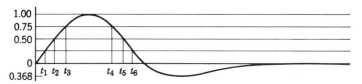

Fig. 2-9 Doubly differentiated pulse divided into increments of equal amplitude to estimate the sum spectrum.

The method is based on the fact that the probability of some critical point on a pulse occurring within a specified small time interval is numerically equal to the width of that interval divided by the mean spacing between pulses. The mean spacing is the reciprocal of the count rate. We begin by graphing the pulse shape obtained from an oscilloscope; we then divide the pulse into a convenient number of equal amplitude intervals. The divided pulse shape for a network consisting of a double RC differentiator followed by a single RC integrator is shown in Figure 2-9. Assume a count rate of N cps. Referring to Figure 2-9, the probability that the peak of the next pulse will fall within the time interval $t_4 - t_3$ is $N(t_4 - t_3)$. This peak will add to the existing one to produce a point on the sum spectrum with a relative amplitude of 1.875 and a relative frequency of occurrence of $N(t_4 - t_3)$. The value of 1.875 ($= 1 + 0.75 + 0.25/2$) was obtained by assuming that the average peak height within the interval $t_4 - t_3$ is at the middle of the amplitude increment. The relative frequency of occurrence for a sum peak height of 1.625 ($= 1 + 0.5 + 0.125$) is $N[(t_3 - t_2) + (t_5 - t_4)]$, for a sum peak height of 1.375 it is $N[(t_2 - t_1) + (t_6 - t_5)]$, etc. Figure 2-10 shows the sum spectrum for a count rate of N monoenergetic pulses per second, each of which has a shape like that given in Figure 2-9.

In Figure 2-10 the main peak occurs at 1.00 on the E (energy) axis. The lower edge of the spectrum occurs at 0.632 and it corresponds to those pulses falling at the trough of the undershoot. The upper edge falls at 2.00, which corresponds to the perfect overlap of two successive pulses. The spectrum of Figure 2-10 is a doubles spectrum, corresponding to the probability of two pulses piling up. The sum spectrum cor-

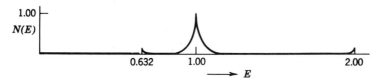

Fig. 2-10 Sum spectrum (doubles) resulting from pulse shape of Figure 2-9.

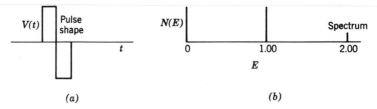

(a)

(b)

Fig. 2-11 (a) Idealized doubly differentiated pulse shape and (b) the resulting sum spectrum.

responding to that of idealized double delay-line differentiation is shown in Figure 2-11.

Overload

The large scintillation detector has a tremendous range of energy response, and thus its largest output signal may be more than a thousand times as large as the smallest signal of interest. The large signals will, of course, drive the amplifier to saturation. If the large pulses lead to extensive amplifier paralysis followed by a slow recovery to the quiescent level, smaller pulses occurring during the recovery interval will not be measured correctly and those occurring during the paralysis interval will be lost altogether. The primary cause of overload paralysis and slow recovery is grid conduction during the positive part of the overload signals. In Figure 2-12, grid current in V_2 rapidly charges the interstage

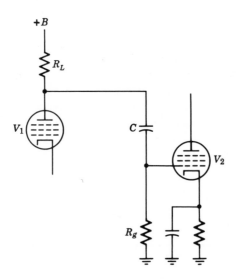

Fig. 2-12 *RC* interstage coupling network.

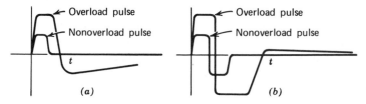

Fig. 2-13 Overload response of delay-line differentiated amplifier: (*a*) before and (*b*) after second differentiator.

coupling capacitor C through the low impedance of the forward-biased grid-cathode diode and the plate load resistor R_L. After the signal decays, the coupling capacitor must discharge through the grid-leak resistor R_g, and this may be many times larger than the charging resistance. The discharge interval, during which the amplifier reference base line is displaced from its quiescent value, may be very much longer than the duration of the overload pulse.

To prevent these overload difficulties we may choose one of the following techniques.

1. Use very small anode load resistors so that the largest possible signal produced at the anode when the tube is driven to cutoff is smaller than the grid bias of the succeeding stage.

2. Make the anode load resistors very large and the grid-leak resistors much smaller.

3. Use catching diodes at the anodes to limit the positive voltage swing to values less than the grid bias of the succeeding stages.

4. Use long-tailed-pair amplifier stages (LTP) instead of pentodes.

5. Introduce double delay-line differentiation in conjunction with circuitry, which keeps the impedance seen by coupling capacitors constant to 10 percent, regardless of the degree of overload. Methods 4 and 5 are commonly used.

Figure 2-13 shows the pulse shapes resulting from the last technique. The good-aspect ratio[1] of the delay-line pulse shapes results in a minimum of primary-pulse broadening, and the second differentiator quickly restores the large undershoot which follows the singly differentiated signal. In the first amplifier to use this technique, recovery to 10 percent from a 400 × overload took about eight nonoverloaded pulse widths and recovery to 1 percent from the same overload took about 100 nonoverloaded pulse widths. The much greater recovery time for the lower percentage recovery is due mainly to the residual "wiggles" caused by nonuniformities in the delay lines. The broadening of the undershoot

[1]The aspect ratio is defined as the width of a pulse measured at 90-percent height divided by the width at 10-percent height.

when in overload is caused by saturation of the amplifier section following the second differentiator; it is undesirable because when the area balance between the primary pulse and the undershoot is upset, a small overshoot is formed which decays with the secondary time constant. In this regard it helps to limit the signal applied to the second differentiator to a level which does not permit it to be overloaded heavily; a typical upper limit is about 4 ×.

Noise

AMPLIFIER SYSTEM NOISE.   Random noise fluctuations in the detector and preamplifier introduce a dispersion which will broaden the pulse-height spectrum resulting from monoenergetic particles. Detector noise is customarily specified in terms of the FWHM (full width at half maximum) broadening of a monoenergetic peak. Since the detector and pre-amplifier are separate and independent sources of noise, the total system noise is equal to the square root of the sum of the squares of the two individual noise contributions.

The lowest noise level attained in the input stage of the preamplifier is an rms signal equivalent to the charge contained in a pulse of 120 electrons (830-eV FWHM referred to a germanium detector). If the detector is capable of 0.25-percent resolution (FWHM), preamplifier noise will control the line width for energies below about 400 keV.

The tolerable noise level of the main amplifier, referred to its input, depends on the preamplifier noise level referred to the preamplifier output terminal. In the example of a commercial preamplifier used with a refrigerated germanium detector, the specifications are as follows:

NOISE LEVEL.   380 electrons rms with a 10-pF detector and 0.8-μsec time constants in the main amplifier.

CONVERSION

GAIN.   0.21 μV per electron, × 1, × 8.

The noise level referred to the input of the main amplifier (or the output of the preamplifier) is 0.21 μV/electron × 380 electrons = 80 μV (or 8 × 80 μV in the high-sensitivity position). The commercially available main amplifier normally contributes no more than 2 percent to the total noise, or no more than 16 μV $[= \sqrt{80^2 - (80 \times 0.98)^2}\,]$ when the preamplifier output noise is 80 μV. If the preamplifier drives a matched cable, the sensitivity is reduced by half and the noise level due to the preamplifier measured at the main amplifier input terminal is halved, but the noise generated in the main amplifier remains unchanged. The principal source of main amplifier noise is the shot effect in the input transistors. Balanced-input stages are normally 41 percent noisier than single-ended stages for the same transistor current.

In the main amplifier, a low-impedance resistive attenuator is generally used as a type of gain control, and it is frequently split into two sections, one at the input and the second at a point later in the amplifier. With this arrangement, the main amplifier noise contribution can be minimized by keeping the input control at the lowest attenuation possible without overloading the input stage.

NOISE SOURCES ASSOCIATED WITH THE PREAMPLIFIER-DETECTOR COMBINATION. The preamplifier-detector combination has the following three types of noise sources, each of which distributes its energy as a function of frequency f.

1. Mean-squared voltage per cycle of bandwidth proportional to $1/f^2$. This type of source includes grid (base) current, thermally generated hole-electron pairs in semiconductor detectors, and thermal noise from the parallel combination of all resistors shunting the preamplifier input terminal to ground.

2. Mean-squared voltage per cycle proportional to $1/f$. This type of source includes noise due to surface leakage current in semiconductor detectors, current noise in composition resistors connected to the preamplifier input terminal, and positive ion emission from the input-tube cathode.

3. Mean-squared voltage per cycle constant with frequency. This type of source includes shot noise from the plate (collector) current of the preamplifier and main amplifier input stages, and thermal noise from damping and feedback resistors connected in series with the grid or cathode of the preamplifier input tube.

INTERACTION OF SHAPING NETWORKS WITH ABOVE NOISE SOURCES. Increasing the resolving time T_W, while keeping the ratio of rise and fall times to T_W constant, increases the $1/f^2$ equivalent rms noise directly as $(T_W)^{1/2}$, does not affect the $1/f$ equivalent rms noise charge, and decreases the constant-f equivalent noise as $\sim 1/(T_W)^{1/2}$.

Holding the resolving time constant but increasing the ratio of high to low frequencies by decreasing the rise and fall times has little effect on the $1/f^2$ noise, increases the $1/f$ noise, and increases the constant-f noise.

Pulse-shaping considerations dictate that the effective input resistance R of the preamplifier be always large compared to the reactance of the input capacitance C in the frequency range passed by the amplifier. The voltage-divider action of this RC network causes the net noise voltage at the amplifier input to be proportional to $1/\sqrt{R}$, so that the thermal resistor noise can be reduced to any extent desired by increasing the effective input resistance.

GROUND-LOOP NOISE. It is difficult to prevent ground-current noise pickup when the preamplifier and main amplifier are separated by 100 ft or more. One way to reduce the ground-loop noise is to use a differential amplifier input stage and a twin axial cable, connected as in Figure 2-14.

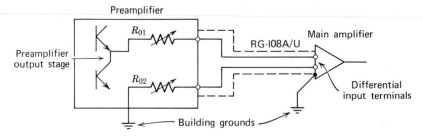

Fig. 2-14   Twin axial cable connected between preamplifier and differential amplifier input stage to reduce ground-loop noise pickup 50 to 200 times with 1-μsec pulse shaping.

One of the cable inputs receives signal from the preamplifier output stage through an adjustable impedance matching resistor. The other cable end is grounded to the preamplifier case, also through an adjustable matching resistor. The desired signal is unbalanced with regard to the differential input and it receives normal amplification. The noise signal, because it enters through the outer shield, furnishes two signals to the main amplifier that are balanced in amplitude and of the same polarity; these signals cancel each other. The resistor R_{01} is adjusted for minimum reflection of the desired signal; R_{02} is then adjusted for minimum noise pickup.

Effects of Temperature on Line Width

The principal effect of temperature on line width is an indirect one; it is caused by the dependence of amplifier gain on temperature. Gain drift over a several-hour run will spread spectral lines just as surely as high-frequency noise. With enough feedback, the overall dependence of amplifier gain variations in transistor parameters can be made negligible. The remaining gain variations then depend on the temperature stability of the resistors and capacitors used in the feedback networks, the gain controls, the input and output terminating resistors, and, if delay-line pulse shaping is used, the delay lines. Components suitable for temperature-stable operation are listed in Table 2-1. In gain controls, pulse-shaping networks, etc., ratios of component values are critical. The effects of temperature variation in different components tend to cancel each other. The degree to which one can rely on this cancellation depends on the temperature gradients within the instrument cabinet and the relative thermal masses of the separate components. The power supply, with its large amount of power dissipation, is a prime generator of thermal gradients.

Table 2-1

Temperature-Stable Components, Suitable
for Critical Circuits (Pulse-shaping Networks,
Gain Controls, and Terminating Resistors).

	Temperature Stability, ppm/°C
Resistors	
Deposited carbon	200 − 400 (negative)
Metal film T-0	±150
″ ″ T-1	±100
″ ″ T-2	± 50
″ ″ T-9	± 25
Wirewound, power, nichrome	+150
Wirewound, precision	± 20
Carbon composition, variable	Not suitable except as a rough trimmer
Cermet, variable	200 − 400 (negative)
Capacitors	
Ceramic disc, NPO	± 30
Ceramic disc, GA, GMV, or HK	Not suitable
Mica, silvered, 100pf and up	0 to +70
Polycarbonate film	− 25 to +140
Polyester film	+100 to +400
TFE film	−200
Polystyrene film	−125
Paper or Oil	Not suitable
Tantalum	Not suitable
Delay lines	
Continuous cables	
Delay	250 − 400
D-c resistance	+4,000
Miniature encapsulated	
Delay	10 − 1,000
D-c resistance	+4,000

2-3 BASIC PREAMPLIFIERS

Cathode- (Emitter-) Follower Preamplifiers

The cathode follower is an amplifier of common-plate configuration,
as in Figure 2-15. Since no bypass capacitor is across the load resistance
in the cathode circuit, constant-current negative feedback is produced,
and the output voltage between the cathode and ground is applied to the
input in opposite phase to the signal applied to the tube. The low-

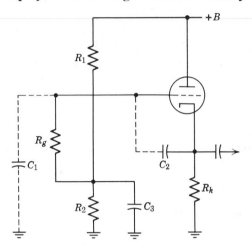

Fig. 2-15   A stable cathode-follower circuit.

frequency voltage gain G of a triode cathode follower is always less than unity and is given by the following expressions.

For the general case

$$G = \frac{\mu R_L}{r_p + R_L (\mu + 1)}. \tag{2-10}$$

For $\mu \gg 1$

$$G \simeq \frac{g_m R_L}{g_m R_L + 1}. \tag{2-11}$$

In the above expressions, μ is the amplification factor, R_L is the net cathode load resistance, r_p is the dynamic plate resistance, and g_m ($= \mu/r_p$) is the transconductance.

When the circuit of Figure 2-15 is used as a preamplifier, the total detector charge Q is made up of two parts, Q_1 on C_1 and Q_2 on C_2. The input voltage is $V_1 = Q_1/C_1$ and the output voltage is $V_0 = GV_1$. The voltage across C_2 is $V_2 = V_1 - V_0 = (1 - G)V_1$. Hence $Q_2 = C_2 V_2 = C_2(1-G)V_1 = C_2(1-G)Q_1/C_1$, or

$$\frac{Q_2}{Q_1} = (1 - G) \frac{C_2}{C_1}. \tag{2-12}$$

Substituting from (2-11) into (2-12), we obtain

$$\frac{Q_2}{Q_1} = \frac{C_2}{C_1 (g_m R_L + 1)}. \tag{2-13}$$

C_1 and C_2 are practically comparable in value, so that if $g_m R_L \gg 1$, Q_2 is negligible compared with Q_1. However, if $g_m R_L$ is not large compared to 1, both G and V_1 are dependent on g_m. Thus, if the charge-to-voltage conversion factor is to be stable, it is necessary to keep g_m constant.

In the cathode follower of Figure 2-15 the effective charge integrating capacitance C_i is equal to $C_1 + C_2 (1 - G)$. The integrating time constant is determined by $R_i C_i$, where R_i is the effective input resistance and is the parallel combination of the detector anode resistor and the grid-leak resistor R_g. The bypass capacitor C_3 keeps noise associated with statistical fluctuations in the voltage-divider current out of the signal circuit. In addition to having a large dynamic signal range (because of the large voltage drop in R_k), the cathode follower also has superior gain stability, since grid-cathode contact potential changes are small compared to the drop in the large cathode resistor.

Figure 2-16 shows a typical example of the cathode follower used as a photomultiplier preamplifier for the scintillation detector.

Figure 2-17 shows the White cathode follower in conjunction with the photomultiplier tube. This circuit has the advantage of having lower output resistance than the preamplifier of Figure 2-16 (or 2-15), combined with the ability to deliver large negative signals to low-impedance loads. As the current falls in the upper tube, a positive signal appears across R_L which causes the lower tube to conduct heavily, thus making a large additional load current available. The output resistance is equal to $1/(g_{m1}g_{m2}R_L)$ and, with ordinary receiving tubes, may be as low as 10 to 20 ohms.

Figure 2-18 shows the Darlington emitter-follower preamplifier used to drive a long cable for less attenuation. This circuit consists of a cascade of three emitter followers in conjunction with a photomultiplier. The emitter resistor R_{e2} (= 47 K) is between R_{e3} (= 470 ohms) and R_{e1} (= 470 K) in value. Since R_{e3} is small compared to R_{e1}, the quiescent current in each stage becomes drastically less as the input of the chain is approached.

Figure 2-19 shows a typical preamplifier for a BF_3 counting system. It also can be used for usual types of proportional counters and scintillation detectors with photomultiplier tubes. It has a voltage gain of 12. It was designed for low noise input and for low-impedance output with adequate gain. An output impedance of less than 10 Ω is suitable to drive a main amplifier with a 93-Ω input. In this circuit the first section is a White source follower (Q_1 and Q_2) which provides low noise input; the output section is a White emitter follower (Q_6 and Q_7), which is used as a low-impedance driver ($Z_0 < 10 \Omega$); between these two sections is a negative feedback loop (see Sec. 2-4) with voltage gain equal to $(R_f + R_k)/R_k$ or about 16 [$\approx (4.99 \times 10^3 + 330)/330$]. This loop consists of two complementary amplifying stages (Q_3 and Q_4) and an emitter follower (Q_5), with inverse feedback from latter to the emitter of the first stage (see Sec. 2-6). The White emitter or source follower operates in the same manner as the White cathode follower described previously.

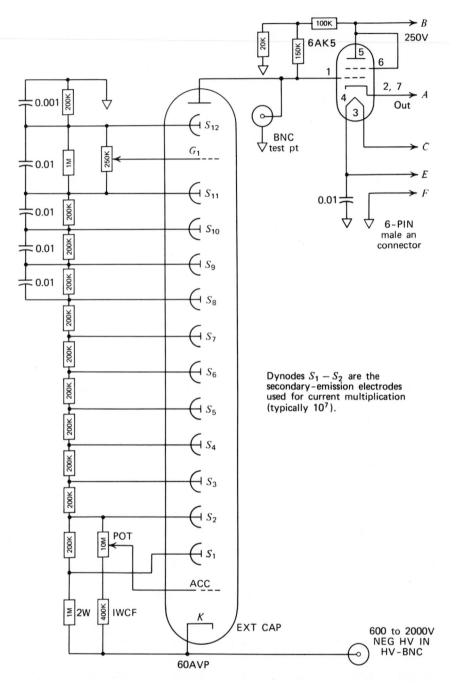

Fig. 2-16 The cathode follower used as the Amperex 60AVP photomultiplier preampli-fier for a scintillation detector. (Model No. PA-114, designed by W. C. Kaiser, Argonne National Laboratory, 1965.)

176

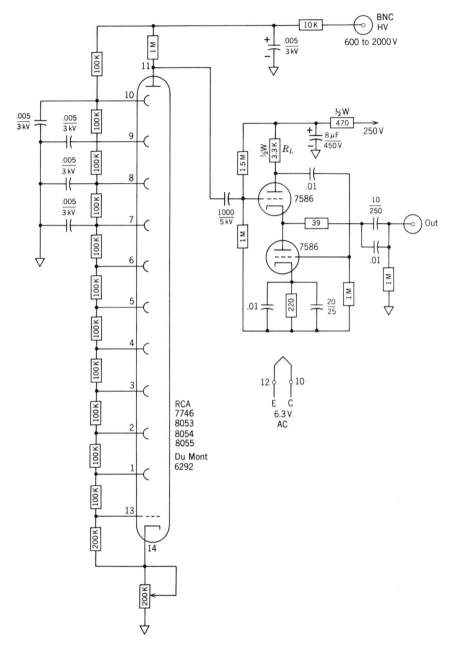

Fig. 2-17   A typical example of the White cathode-follower preamplifier in conjunction with a photomultiplier. (Model PA-129, designed by G. Ansley, EL Div., Argonne National Laboratory, 1966.)

177

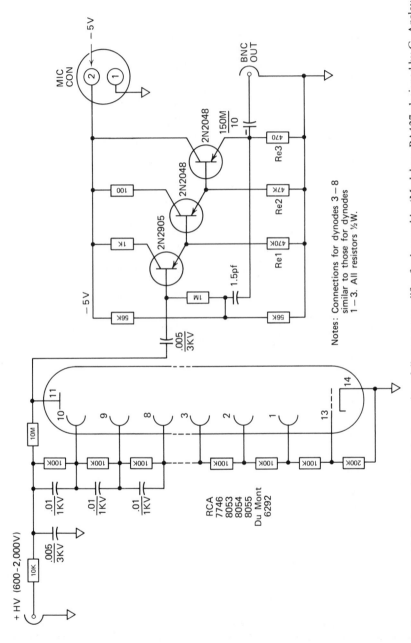

Fig. 2-18 Photomultiplier base with Darlington emitter-follower preamplifier for long cable. (Model no. PA127, designed by G. Ansley, EL Div., Argonne National Laboratory, 1966.)

178

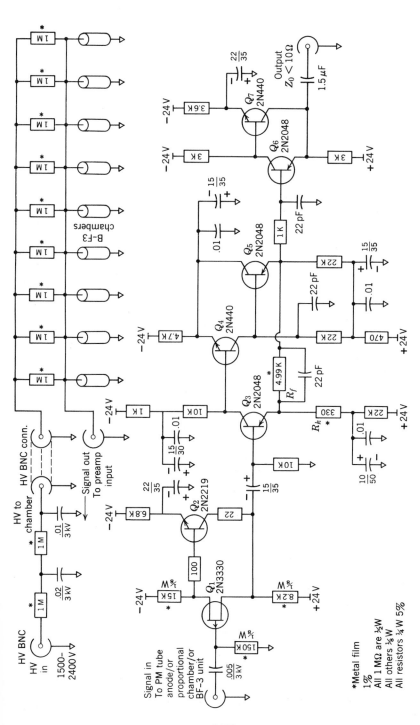

Fig. 2-19 Preamplifier for BF$_3$ counting system. (Model no. PA147, designed by Strauss and Miranda, EL Div., ANL, 1966.)

179

Integrator Preamplifiers

Figure 2-20 shows an integrator preamplifier consisting of two transistors Q_1 and Q_2. The signal charge Q divides into two parts, Q_i on C_i and Q_f on C_f. The input voltage is $V_i = Q_i/C_i$ and the voltage at the emitter of Q_2 is $V_{e0} = -GV_i$, where G is the voltage gain of Q_1. The voltage across the feedback capacitor C_f is $V_i - V_{e0}$, so that $Q_f = C_f(V_i - V_{e0}) = C_f V_i(1+G) = C_f Q_i(1+G)/C_i$, or

$$\frac{Q_f}{Q_i} = \frac{C_f(1+G)}{C_i}. \tag{2-14}$$

Thus if C_f and C_i are comparable and G is large, Q_f is large compared with Q_i and $Q \simeq Q_f$. Furthermore, if G is large, V_{e0} is large compared to V_i, and $V_{e0} \simeq Q_f/C_f \simeq Q/C_f$. The voltage V_{e0} at the emitter of Q_2 can be used as the preamplifier output to drive the next amplifier stage provided that it is a high-input-impedance voltage-sensitive stage. More commonly, however, the next stage is a low-input-impedance current-sensitive stage, so the preamplifier output is usually taken from the collector of Q_2. Note that there is a "bootstrap" connection between the emitter of Q_2 and the collector load resistor of Q_1 through capacitor C_2. This connection increases the effective collector load resistance of Q_1 and, hence, increases the feedback-loop internal gain. The effective resistance in the emitter circuit of Q_2 is, therefore, equal to $R_{e0} = R_1 R_2/(R_1 + R_2) \simeq R_2$ (because of $R_2 \ll R_1$). Neglecting the base current of

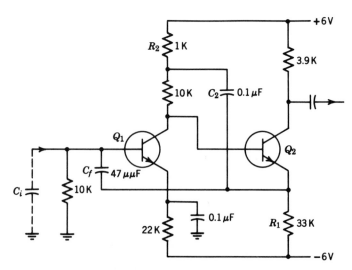

Fig. 2-20   Integrator preamplifier. (From F. Goulding, *IRE Trans. Nucl. Sci.,* NS-5, p. 38, 1958.)

Q_2, the collector current (i_0) of Q_2 is approximately equal to the emitter voltage V_{e0} divided by the emitter circuit resistance R_{e0}, or

$$i_0 \simeq \frac{V_{e0}}{R_{e0}} \simeq \frac{Q}{R_2 C_f}. \tag{2-15}$$

2-4   FEEDBACK LOOPS

Linear pulse amplifiers are always made up of one or more negative-feedback loops. Figure 2-21 shows the block diagram of a negative-feedback loop, in which an amplifier having a gain of K without feedback has a fraction β of its output voltage V_0 fed back to add out of phase with the signal voltage V_s. The net input voltage is $V_i = V_s - \beta V_0$. Also $V_i = V_0/K$. Hence $KV_s = V_0 (1 + \beta K)$. G, the gain with feedback, is then

$$G = \frac{V_0}{V_s} = \frac{K}{1 + \beta K}. \tag{2-16}$$

If K is sufficiently large so that $\beta K \gg 1$, then

$$G \simeq \frac{1}{\beta}. \tag{2-17}$$

The feedback factor β is determined by passive linear-circuit elements. If the amplifier has a large βK product (reserve gain), the gain with feedback is seen to depend only on the linear and stable feedback factor β and is independent of the relatively unstable and often nonlinear values of transconductance of tubes and transistors. Thus stability and linearity are improved. However, at the extremes of the frequency bandpass, the βK product is not large compared to 1, so that the stability and linearity for signals having either high- or low-frequency components are not significantly improved by negative feedback. Hence the ideal linear amplifier has adequate bandwidth for the input-signal frequency spectrum before negative feedback is applied.

A two-transistor feedback loop is shown in Figure 2-22. It is a current-sensitive circuit which employs current feedback from the emitter of Q_2 to the base of Q_1 through R_f. Neglecting the base current in Q_2, the output current is approximately equal to the emitter current of Q_2, and so the current feedback factor is that fraction of the emitter current that flows in R_f. The net resistance in the emitter circuit of Q_2 is $R_{e0} = 1/(1/R_3 + 1/R_f + 1/R_1)$. For the values shown in the circuit, R_{e0} is approximately equal to R_1. Therefore the current feedback factor may be given by

$$\beta \simeq \frac{R_1}{R_f}. \tag{2-18}$$

Since the bootstrap connection (via the 30-μF capacitor) increases the effective collector load resistance of Q_1, nearly all of the signal current from Q_1 flows in the base circuit of Q_2. Hence the unfedback current

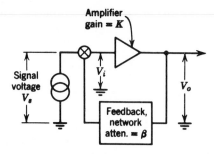

Fig. 2-21   Block diagram of a negative-feedback loop.

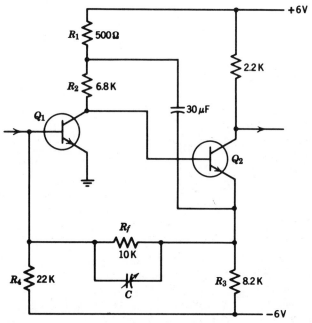

Fig. 2-22   Current-amplifying feedback loop consisting of two-transistor stages. (From F. Goulding, *IRE Trans. Nucl. Sci.*, NS-5, p. 38, 1958.)

gain K is the product of the current transfer ratios of the two transistors. If the βK product is large compared to 1, the gain with feedback is $1/\beta$. The trimmer capacitor C is used to control the high-frequency oscillatory tendencies due to stray capacitance to ground at the collectors, etc. This capacitor is usually adjusted empirically to that value which results in a minimum output-signal rise time without overshoot.

2-5 LOW-NOISE PREAMPLIFIERS FOR SEMICONDUCTOR RADIATION DETECTORS

Charge-Sensitive Circuits

The charge-sensitive circuit in Figure 2-23 has become widely accepted as the optimum preamplifier for use with semiconductor radiation detectors. This has occurred primarily because the effects on pulse height of variations in the impedance of the semiconductor detector are minimized by the low-input impedance of the charge-sensitive preamplifier. However, when this amplifier acts as a charge integrator, the output pulse is observed to cross the baseline. Furthermore, the bipolar characteristic of this pulse is of great practical significance because it has been observed that under overload conditions the undershoot of the nominally unipolar pulse significantly increases the dead time during which the amplifier is unable to transmit information. Therefore the undesirable undershoot must be eliminated. This problem is usually solved by means of pole-zero[1] cancellation, as shown in Figure 2-26.

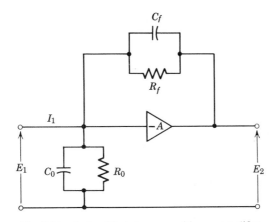

Fig. 2-23   A simplified charge sensitive preamplifier.

[1]The definitions of pole and zero are related to network functions. All network functions have the form of a quotient of polynomials as $(a_0 s^n + a_1 s^{n-1} + \cdots + a_{n-1} s + a_n)/(b_0 s^m + b_1 s^{m-1} + \cdots + b_{m-1} s + b_m)$. If the numerator polynomial is factored into its n roots, and the denominator polynomial is factored into its m roots, the equation can be written in the form $H(s - s_1)(s - s_2) \cdots (s - s_n)/(s - s_a)(s - s_b) \cdots (s - s_m)$, where $H = a_0/b_0$ is a constant known as the scale factor and the roots $s_1, s_2, \ldots, s_a, s_b, \ldots$ are complex frequencies. When the variable s has the values $s_1, s_2, \ldots, s_n$, the network function vanishes. Such complex frequencies are called zeros of the network function. When s has the values $s_a, s_b, \ldots, s_m$, the network function becomes infinite. These complex frequencies are called poles of the network function. The complex variable $s = \sigma + j\omega$ is the Laplace transform variable. The Laplace transform is given by $F(s) = \int_0^\infty f(t)\epsilon^{-st} \, dt$.

The design of present charge-sensitive preamplifiers is frequently based on the charge integrator employing a cascode pair with dynamic (bootstrapped) load. A typical configuration incorporating the most important features is shown in Figure 2-24. The input stage can be an electron tube, a bipolar transistor,[1] or a field-effect transistor. The simplest circuits omit the first emitter follower (Q_3). The principal advantages of this circuit are that it has negligible Miller effect in the first stage and high gain of the cascode pair and wide dynamic range. A parameter characterizing this circuit is the "reserve gain," which is defined as

$$G_r = \frac{G_0 C_f}{C_{\text{in}}}, \tag{2-19}$$

where G_0 is open-loop gain, C_f is the feedback capacitance, and C_{in} is the total input capacitance without feedback effects. The open-loop gain is determined by the product of input-stage transconductance g_m and the dynamic resistance R_d:

$$G_0 = g_m R_d. \tag{2-20}$$

The dominant open-loop time constant (τ_0) is determined by the dynamic resistance and the stray capacitance (C_s) at the output of the cascode,

$$\tau_0 = R_d C_s. \tag{2-21}$$

The closed-loop time constant τ_c is equal to that of the open-loop time constant divided by the reserve gain,

$$\tau_c = \frac{\tau_0}{G_r} = \frac{C_s}{g_m} \frac{C_{\text{in}}}{C_f} \tag{2-22}$$

and is independent of dynamic resistance. This means that an increase in load resistance—by more effective bootstrapping, for example—increases the reserve gain without affecting preamplifier response time. Provision is sometimes made for adjustment of operating conditions, which is necessary for noise optimisation.

Electron tubes employed in the input stages are superior at room temperature. Bipolar transistors are useful in all applications where low noise is not important. Field-effect transistors (FET's) give the best noise performance at low temperatures; the noise levels, known to be a minimum in the region from 110 to 130°K, can be explained in terms of two factors. First, the channel noise at low temperatures consists basically of two components: thermal noise, which increases with temperature, and the fluctuations in number of activated carriers ("carrier-activation" noise), which decreases with temperature. Second, the transconductance has a maximum in the same temperature region. This results from its depen-

[1]A bipolar transistor is a bipolar device whose performance depends on the interaction of two types of charge carriers, holes, and electrons. The field-effect transistor is a unipolar device (see Sec. 1-11).

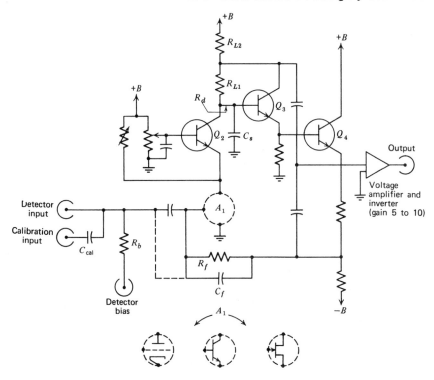

Fig. 2-24 Charge-sensitive preamplifier employing cascode pair (A_1 and Q_2) and dynamic ("bootstrapped") load resistance (R_d). Cascode arrangement provides high loop gain. [From V. Radeka, Nucleonics 23: 7, p. 55, (July 1965).]

dence on the number of activated carriers, which increases with temperature, and on the carrier mobility, which starts to decrease with temperature in this region.

The Typical Preamplifier for Semiconductor Detectors

Figure 2-25 shows the circuit diagram of the typical preamplifier for semiconductor detectors. The charge-sensitive section has very high internal gain—over 60,000 at room temperature. Direct coupling was used to simplify count-rate problems. Though the first two transistors (Q_0 and Q_1) form a cascode, this configuration was chosen for overall design considerations. A germanium microalloy transistor (2N2402) is used after the input FET (Q_0) because it exhibits good gain-bandwidth characteristics as well as low noise. The current in this transistor is only 0.2 mA. The cascode gain is almost 500 at room temperature. A source follower (Q_2 = 2N3823) is used to avoid loading the cascode while providing high gain-bandwidth product. Relatively noisy FET's can be used

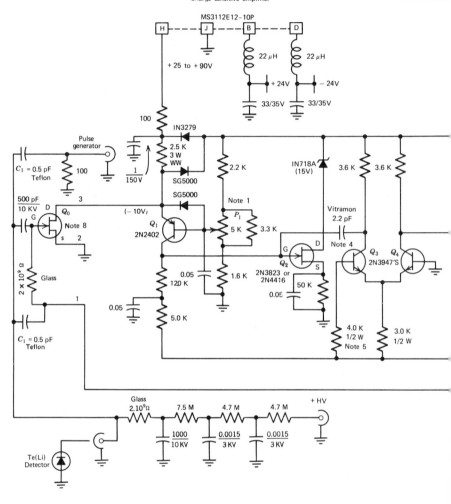

Charge sensitive amplifier

Fig. 2-25 Room temperature FET semiconductor detector preamplifier (Model no. PASD-4A, designed by I. S. Sherman, EL Div., Argonne National Laboratory, 1968.)

Notes

1. P_1 is adjusted for desired FET voltage, typically + 10V at ③ .
2. P_2 is adjusted for pole-zero compensation.
3. P_3 is adjusted for zero DC at the output.

186

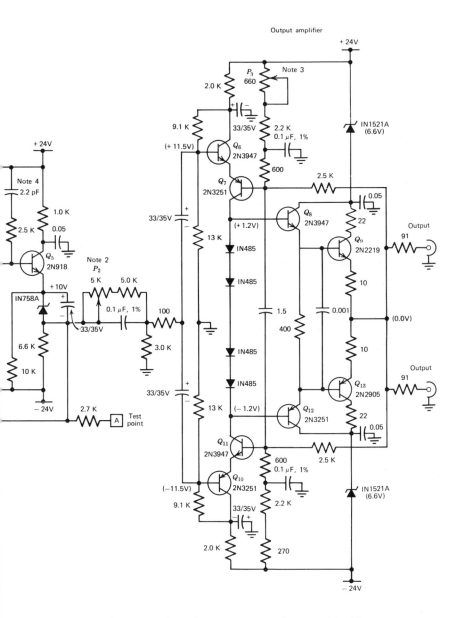

Output amplifier

4. These two capacitor values depend on detector capacitance and feedthru arrangement.
5. This resistor is selected for zero to a few tenths of a volt, negative gate-source bias on Q_2.
6. Resistors are ⅛ W, metal-film except where noted.
7. Capacitors are in microfarads except where noted.
8. Depending on desired characteristics, Q_0 can be selected from a number of types: 2N4416, 2N3823 (or SF5868), TIS88, TIS75, and 2N4393.

187

at this point. The gain of the source follower is 0.8 and the gain of the difference amplifier (Q_3 and Q_4) is 200. A fast emitter follower (Q_5) drives the capacitance of the feedback-loop wiring and the output amplifier.

The output amplifier is capable of driving two 91-ohm terminated lines to about 4.5 volts before overloading. It has a rise time of 11 nsec for either positive or negative pulses. Its closed-loop gain up to the source terminating resistors is 5, and its open-loop gain is about 2500. The last four transistors (Q_8, Q_9, Q_{12}, Q_{13}) are already well degenerated, and the overall gain stability is high. The total base-emitter capacitance at the point in the circuit where the diode bias string is located is almost 20 pF. For an 11-nsec rise-time pulse at full output the peak charging current is about 20 mA. Depending on the pulse polarity, one side of the push-pull input stage provides this current. The quiescent current is 1 mA; a low current is desirable for low base current noise. After this capacitance is charged, the input stage returns to more linear class A operation.[1]

The output amplifier feedback loop and the RC coupling network (see Figure 2-26) provide pole-zero cancellation to eliminate the undesirable undershoot on the output pulse. As shown in Figure 2-26, the capacitor C_2 is chosen to produce a pole having a 50-μsec time constant, the value desired for the final output tail pulse. The resulting zero, whose time constant is 110 μsec, is canceled by the coupling network (R_1-R_2-C_1). This

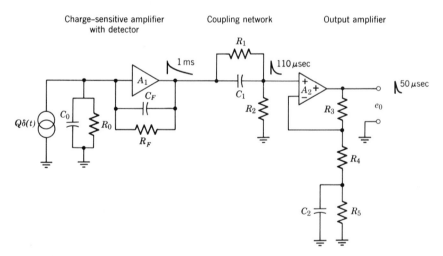

Fig. 2-26   Simplified form of the preamplifier (PASD-4A) shown in Figure 2-25. The coupling network is an RC differentiator with pole-zero cancellation.

[1]A class A amplifier is an amplifier in which the base bias and alternating signal are such that the collector current in a transistor flows continuously during the complete electrical cycle of the signal, and even when no signal is present.

network has a 110-μsec pole (canceling the 110-μsec zero) and a zero at 1 msec that cancels the time constant of the charge-sensitive section.

The principal characteristics of the preamplifier (PASD-4A) are listed below:

1. Sensitivity: 0.3 volt/MeV [relative to charge from Ge(Li) detector] terminated with 91 ohms.

2. Maximum signal level: at least 12 MeV.

3. Gain stability: 10 ppm drop in closed loop gain for 1-percent reduction in gain of charge sensitive amplifier with 5-pF detector.

4. Output pulse shape: 50-μsec tail, 20-nsec rise time for low capacitance increasing to 30 nsec for 100 pF.

5. Overload characteristics: 50-μsec exponential recovery of overload pulses up to 68 MeV.

6. Counting rate capability: 2½-keV resolution has been obtained at 50,000 pulses/sec at 1.33 MeV.

7. Noise with input stage at room temperature: 600 to 700 eV at zero capacitance with increase of 50 eV/pF added capacitance.

8. Noise at preamplifier output: less than 15 keV with 40-nsec rise time with a 5-pF Ge(Li) detector.

2-6 TYPICAL CIRCUITS FOR THE MAIN LINEAR AMPLIFIERS

The block diagram and schematic circuit of the low-noise linear pulse-amplifier model LPA-19 are shown in Figures 2-27 and 2-28, respectively. It primarily consists of the input inverter (Q_1 and Q_2), first feedback loop ($Q_3 - Q_5$), second feedback loop ($Q_6 - Q_8$), output driver (Q_9 and Q_{10}), and output inverter (Q_{11} and Q_{12}). In order to improve the count rate capability of the amplifier, a base-line restorer is added between the second feedback loop and the output driver, and an *RC* differentiating network with pole-zero cancellation is inserted between the first and second feedback loops.

The two feedback loops are noninverting amplifying loops with a voltage gain of 16 which is determined by the ratio $(R_3 + R_4)/R_3$ or $(R_9 + R_{10})/R_{10}$ [$= (4.99 \times 10^3 + 330)/330$]. Since virtually no quiescent current flows in the feedback resistors R_4 and R_9, the gain of the loop can easily be changed without altering the dc conditions of the amplifier. The differentiating network has a gain of 0.37 for equal integrating and differentiating time constants so that the total or nominal gain of the amplifier is $16 \times 0.37 \times 16 = 95$. The gain factor may be increased or decreased by 2 or 4 by increasing or decreasing R_4 and/or R_9 by a factor of 2. The total gain is continuously adjustable from zero to the nominal maximum of 95.

As shown in Figure 2-27, τ_2, τ_3, τ_4 are the three time constants related to the signal. $\tau_2 = 50$ μsec nominal for the input pulse tail; $\tau_3 = R_4C_3$

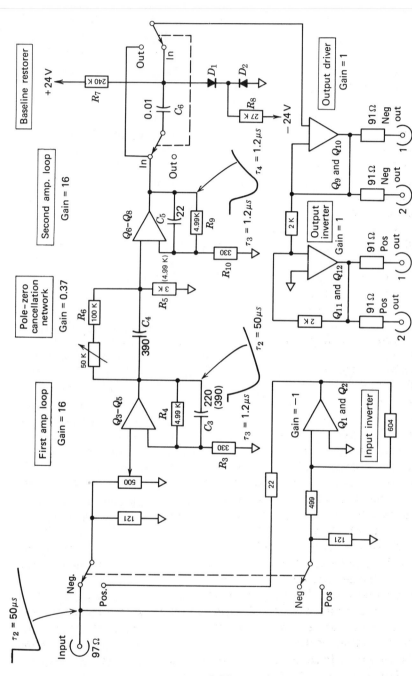

Fig. 2-27 Block diagram of the low noise linear pulse amplifier model LPA-19 (designed by M. Strauss and R. Brenner, EL Div., Argonne National Laboratory, 1967).

190

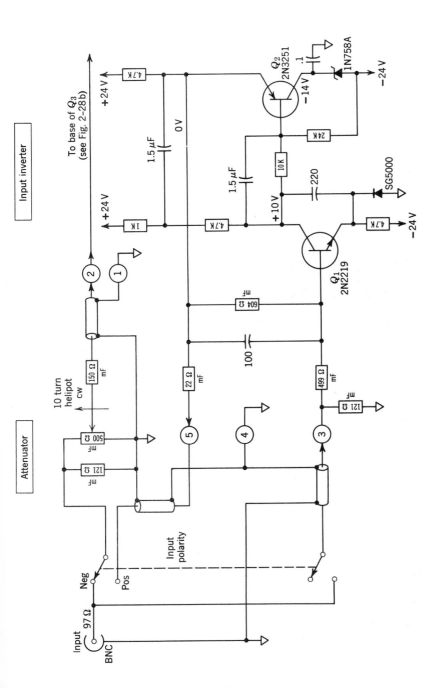

Fig. 2-28a Circuit diagram of attenuator and input inverter of the low noise linear pulse amplifier model LPA-19 (designed by M. Strauss and R. Brenner, EL Div., Argonne National Laboratory, 1967).

191

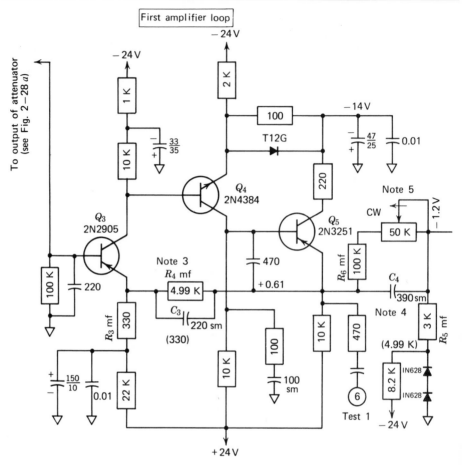

Fig. 2-28*b* Circuit diagram of first and second amplifier loops of the low noise linear pulse amplifier model LPA-19 (designed by M. Strauss and R. Brenner, EL Div., Argonne National Laboratory, 1967).

Notes for Figs. 2-28a-d unless otherwise specified:
1. All resistor ¼ W carbon

$= 4.99 \times 10^3 \times 220 \times 10^{-12} \simeq 1.2$ μsec for integrating; $\tau_4 = R_5 C_4 = 3 \times 10^3 \times 390 \times 10^{-12} \simeq 1.2$ μsec for differentiating. To obtain $\tau_3 = \tau_4 = 2$ μsec, use $C_3 = 390$ pF, $R_4 = 4.99$ K, $C_4 = 390$ pF, and $R_5 = 4.99$ K. Double integration may be accomplished by increasing C_5.

The output inverter provides positive and negative outputs from 0 to 10 V. The outputs are buffered with a series 91-ohm resistor. Four different outputs are available, two negative and two positive, each with its own terminating resistor to prevent cross-talk between them.

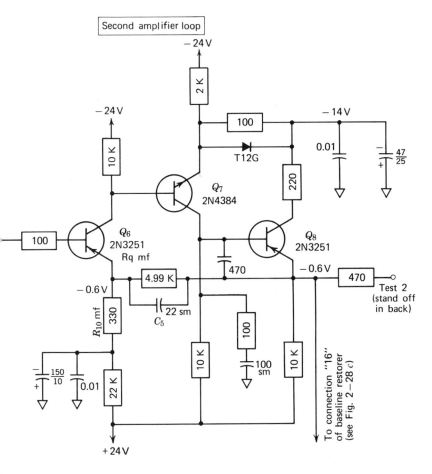

2. Electrolytic cap. dry tantalum
3. Integrating T.C. = R_4C_3 = 1.2 μs or (2 μs)
4. Differentiating T.C. = C_4R_5 = 1.2 μs or (2 μs)
5. Adjust for no undershoot at the test 2 point or at the output (baseline restorer out)
6. Neg #3DC679

The output noise is independent of gain setting and is on the order of 2 mV peak-to-peak. Overloads of up to 100 times full scale do not result in any spurious effects.

Another typical linear pulse-amplifier circuit is shown in Figure 2-29. This amplifier was designed to operate with a preamplifier in which the first RC differentiation (2 μsec) is performed. To limit the low-frequency response, clip the tail of the single differentiated signal and reinstate the base line; a second RC differentiation (R_2C_2 = 8 μsec) and a dc restora-

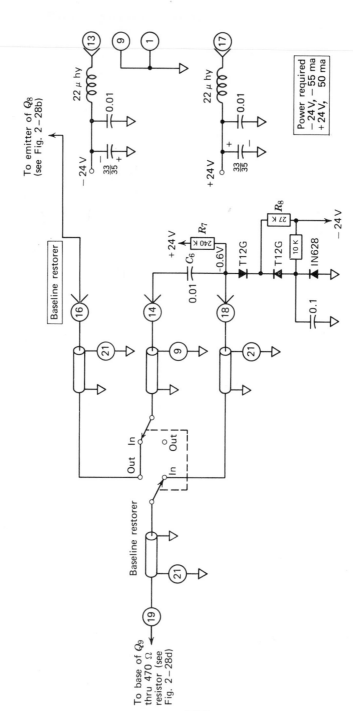

Fig. 2-28c Circuit diagram of baseline restorer of the low noise linear pulse amplifier model LPA-19 (designed by M. Strauss and R. Brenner, EL Div., Argonne National Laboratory).

194

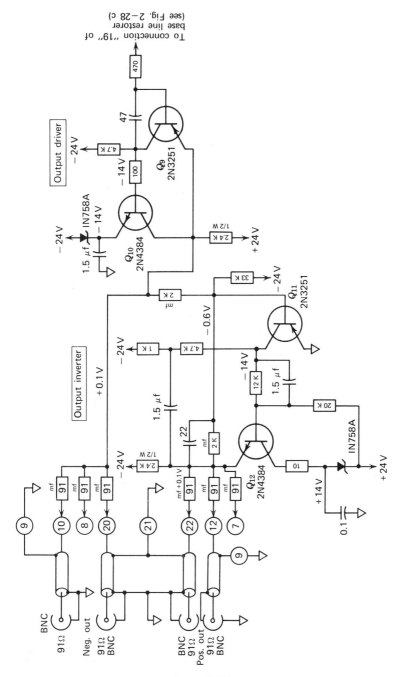

195

Fig. 2-28d   Circuit diagram of output driver and inverter of the low noise linear pulse amplifier model LPA-19 (designed by M. Strauss and R. Brenner, El Div., Argonne National Laboratory, 1967).

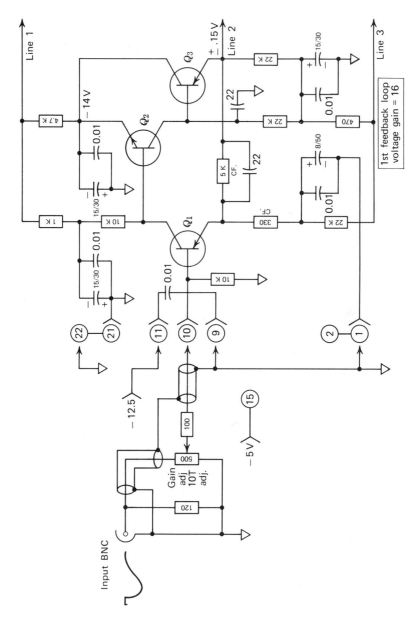

Fig. 2-29a First feedback loop of the linear pulse amplifier model LPA-4 (designed by M. G. Strauss, EL Div., Argonne National Laboratory, 1963).

196

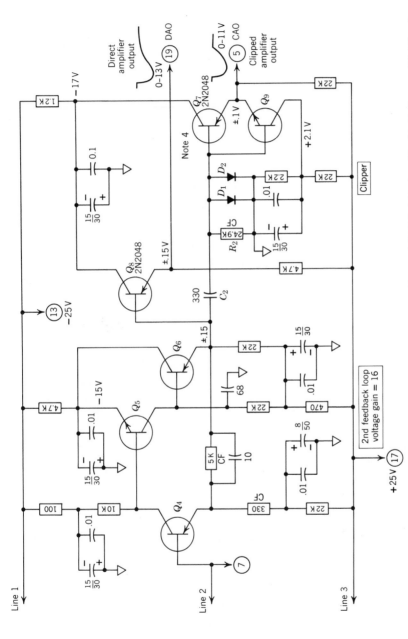

Fig. 2-29b Second feedback loop and output drivers of the linear pulse amplifier model LPA-4 (designed by M. G. Strauss, EL Div., Argonne National Laboratory, 1963).

197

tion (D_1, D_2) are performed at the output of the amplifier (Figure 2-29). The amplifier is noninverting, with a full-scale output of -10 V and a rise time of 0.25 μsec. The voltage gain of about 200 is effectively controlled by means of a 10-turn linear attenuator at the input. The input impedance is about 93 ohms, which is suitable for cable termination.

The amplifier is made of two virtually identical feedback loops. Each loop consists of two complementary amplifying stages and an emitter follower. The closed-loop gain and open-loop gain of each ring are 16 and about 250, respectively. The amplifier has two high-gain output drivers, which provide the direct output and the clipped output. The clipped-amplifier output stage, as originally designed, was used to drive the linear gate which was direct-coupled to the complementary emitter follower (Q_7 and Q_9); therefore a dc-level change at this driver constitutes a base line shift. To minimize such a shift at high count rates, R_2 is made large compared to the forward resistance of the diode restorer (D_1 and D_2) (see Sec. 1-9). The collector-to-base leakage current of Q_7 is supplied by the base of Q_9. If the leakage currents of the two transistors are comparable, the potential across R_2 becomes insensitive to variations in these currents, thereby temperature-stabilizing the base line.

2-7 SINGLE-CHANNEL PULSE-HEIGHT ANALYZERS

Introduction to the Single-channel Analyzer

The height of the output pulse from a linear amplifier connected to a detector is proportional to the energy dissipated by the nuclear radiation within the detector, and so the energy-distribution curves (nuclear spectra) may be obtained by measuring the amplitudes of the output pulses. The measuring instrument for obtaining the pulse-height-distribution curves is known as a pulse-height analyzer. This apparatus is used to determine the total number (or sometimes the rate of arrival) of pulses whose heights fall in selected intervals or channels throughout the range of pulse heights of interest. If this analysis is made by sampling one channel at a time, the apparatus is called a single-channel pulse-height analyzer (SCA). Thus the SCA is capable of recording only those pulses

---

Notes for Figs. 2-29 a & b unless otherwise specified:
1. All resistors $\pm 10\%$ $\frac{1}{2}$ W
   all capacitors $\pm 20\%$
   μF if < 1 or $\mu\mu$F if > 1
2. All diodes T12G
3. All PNP transistors 2N2048
   All NPN transistors 2N440
4. Q7 and Q9 are selected so that their I_{CO} are equal @ $V_{CBO} = 1$ V
5. This DC card must have a ground bus-bar, and be placed between two steel shields
6. Lines 1, 2, 3 in Fig. 2-29a are connected to Lines 1, 2, 3 in Fig. 2-29b, respectively

falling within a single channel or section of an energy spectrum; all other pulses are rejected.

A block diagram of a SCA is shown in Figure 2-30. This arrangement includes two discriminators. They are the lower level discriminator (LLD), with the channel level E, and the upper level discriminator (ULD), with the level $E + \Delta E$. Here E is the threshold setting and ΔE is the channel window or window width, which is a small, usually adjustable difference in the triggering levels of ULD and LLD. These discriminators are followed by the anticoincidence circuit (inhibitor), which is arranged to pass a signal only when a pulse passes the LLD but not the ULD and to reject all signals that simultaneously pass both the LLD and the ULD. Therefore, the SCA generates an output pulse if and only if an input pulse is received which has an amplitude within a selected voltage window ΔE. Thus the counter (scaler) registers only those pulses whose height lies in the interval or window ΔE.

In obtaining a differential spectrum, a series of 100 observations can be made in which each measurement of activity is taken in 1-volt increments for a 1-volt window width over a 100-volt range. Plotting the activity/volt (or count rate) versus the voltage or pulse height (channel level or threshold setting) yields a differential spectrum. These increments are referred to as channels, and 100 channels are measured. Single-channel analyzers are used to make the measurements with a single channel at a time, while multichannel analyzers allow pulses in all channels to be recorded simultaneously.

The single-channel analyzer (Figure 2-30) is operated in a differential mode, and so is a differential pulse-height analyzer. However, some commercial-single channel analyzers may be operated in either a differential or an integral mode. In the integral mode, the analyzer generates an output pulse if and only if an input is received with an amplitude above the selected threshold voltage level.

The RIDL Model 33-10B Anti-walk Single-channel Analyzer

GENERAL. The Model 33-10B single-channel analyzer may be operated in either a differential or an integral mode. The analyzer voltages

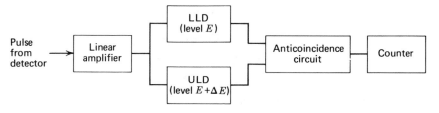

Fig. 2-30 Block diagram of a single-channel analyzer (differential mode).

and corresponding energy pulses are shown in Figure 2-31. Model 33-10B has an output-pulse time stability of less than 10 nsec shift for a 1- to 10-V double delay-line input pulse with a 0.25-μsec or less rise time with the 3-V output. The Analyzer temperature stability is 0.5 mV/°C. The analyzer can be used at high count rates. Rates of the order of 10^6 counts per second may be handled with good pulse-height resolution. Fast, stable coincidence experiments may be performed with the analyzer because of the zero-crossing-type lower-level discriminator. A delayed coincidence system is shown in Figure 2-32, where the Model 33-10B is in one branch, and a Model 34 series multichannel analyzer is in the other branch. The Model 33-10B analyzer itself consists of a lower-level

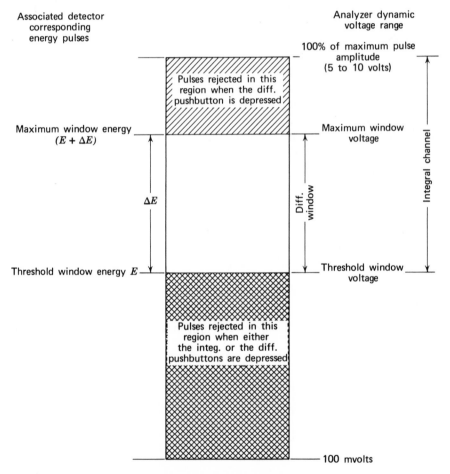

Fig. 2-31   Analyzer voltages and corresponding energy pulses for RIDL Model 33-10B.

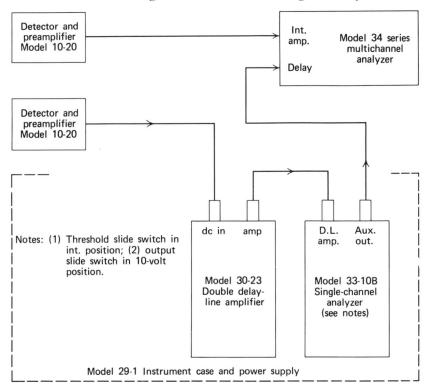

Fig. 2-32   A delayed coincidence system with an RIDL Model 34 Series Multichannel Analyzer.

discriminator and an upper-level discriminator, with the outputs from both discriminators connected in anticoincidence.

INTEGRAL MODE.   In the integral mode, the WINDOW potentiometer and its associated anticoincidence circuit are ineffective. An output pulse is generated only if an input is received with an amplitude above the selected threshold voltage level. The source of the threshold level E is determined by the THRESHOLD slide switch. When the switch is in the INT. position, E is controlled by the setting of the 10-turn precision threshold potentiometer. When the threshold slide switch is in the EXT. position, an external voltage controls the threshold level. This external voltage should be limited to a potential between ground and $+5.6$ V. Each threshold dial division represents 0.1 percent of full scale. Any threshold setting determines the smallest amplitude voltage pulse which will be accepted; this corresponds to the minimum energy of radiation which is accepted through the system from the associated detector. The output pulse is normally used to drive a scaler or a count rate meter. The linearity is 0.25 percent.

DIFFERENTIAL MODE. The variable ΔE energy range is the window of the analyzer in the differential mode, and it is a difference between the threshold level E and a second established level $E + \Delta E$. As in the integral mode, the threshold slide switch permits the threshold potentiometer or some external voltage to set the window's lower level. The window width ΔE is set with the WINDOW potentiometer; this control is identical to the threshold potentiometer with 1000 indexed settings. The window's upper level, $E + \Delta E$, is determined by adding the levels set by each of the two controls.

The maximum output of the associated amplifier will be some specific amplitude less than or equal to 10 V. Any combination of settings of the threshold and window potentiometers which add up to more than this specific voltage will place the upper window level above the maximum voltage level which will be furnished for analysis; this effectively eliminates the upper window level. When the combination of voltages of the two controls yields a sum less than the maximum amplifier output voltage, the analyzer, operated in its differential mode, generates an output pulse if and only if an input pulse is received which has an amplitude within the specified window. In the differential mode, linearity is 0.50 percent or better, and the output pulses are used normally to drive a scaler or a count-rate meter.

The maximum dynamic range of the analyzer is 0.1 to 10 V (see Figure 2-31). When the threshold slide switch is in the INT. position, an adjustable potentiometer (E RANGE) mounted internally permits adjustment of full scale over the range of 5 to 10 V. This potentiometer is in series with the threshold control. The combination of settings of the controls is illustrated in the following example. When the Model 33-10B has been calibrated for a maximum pulse amplitude of 10 V, setting the base line at 5 V (threshold potentiometer at 500 divisions) and the window at 1 V (window potentiometer at 100 divisions) results in a 1-V window between 5 and 6 V.

The analyzer in the differential mode must judge two items of information: first, whether or not the pulse has crossed the lower boundary, and second, whether or not the pulse has crossed the upper boundary. The signals indicating a lower- or upper-level crossing are obtained from two voltage discriminators. If the input pulse triggers the lower-level discriminator but not the upper-level discriminator, a pulse from the lower discriminator passes through an anticoincidence circuit and produces an output pulse. If the input pulse triggers both upper and lower level discriminators, the upper level discriminator output will be sent to the anticoincidence circuit to inhibit the output pulse. The anticoincidence circuit is characterized by its high speed. The analyzer is capable of accepting a new pulse only one microsecond after the termination of a previous pulse. An output pulse is generated in the differential mode only if the lower discriminator is triggered without the coincident triggering of the upper discriminator.

WALK (TIME SHIFT) AND ANTIWALK IN DISCRIMINATOR OUTPUTS. Experiments involving time of flight or coincidence studies require precise and stable time determination of the occurrence of nuclear events of interest. Detection and amplification of these events result in a voltage pulse whose amplitude is proportional to energy loss in the detector and whose rise time is characteristic of the detector energy-to-charge conversion process. This pulse is then examined by a single-channel pulse-height analyzer. The single-channel analyzer is normally used to determine the energy region of interest and the time of occurrence of the event satisfying the energy criterion. Time determination in the analyzer is usually derived from the lower discriminator (typically a Schmitt trigger) which defines lowest voltage or energy level which the analyzer will accept. When the time determination is derived from the leading edge of this discriminator (Figure 2-33), there is a walk (time shift) in the leading edge of the discriminator output due to the rise time of the pulses being analyzed. This walk ΔT will depend on the system rise time and range of pulse amplitudes. Typical resolving times as a result of this walk in coincidence experiments using NaI scintillation detectors are 0.05 to 0.25 μsec.

The walk or time shift ΔT can be eliminated by using the zero-crossing-type lowel-level discriminator in the Model 33-10B analyzer. This discriminator is a Schmitt circuit (a regenerative trigger), which exhibits a characteristic known as hysteresis. The hysteresis of a regenerative trigger is the input-voltage difference between the voltage necessary to trigger the discriminator and the voltage to which the input must drop to allow the discriminator to recover to its quiescent state. The lower discriminator in the Model 33-10B contains a feedback system (Figure 2-35a) which is incorporated into the discriminator level control. When an input pulse triggers the lower discriminator, the feedback system changes the discriminator hysteresis so that the discriminator returns to its normal state as the input pulse crosses the base line, regardless of the discriminator level (Figure 2-34). An output pulse is generated at this time so that the single-channel-analyzer output can also be used to indicate the arrival time of the event. The advantage of this type of circuit is that the lower discriminator can be set at any level without changing the antiwalk characteristics of the unit. Another advantage is that high counting rates at low energies (below the discriminator level) do not affect the circuit, since the antiwalk feature operates only when the

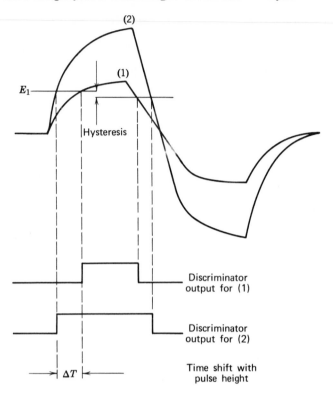

Fig. 2-33   Basic discriminator output pulses for two input pulses.

lower discriminator is triggered. Time resolution in the order of 5 to 10 nsec, independent of discriminator level, is typical for this circuit.

CIRCUITRY OF THE MODEL 33-10B.   The circuitry of the analyzer shown in Figures 2-35a, b, c, and d includes the basic circuit, regulator circuit, output shaper and inverter, and interconnection of the entire system, respectively. In Figure 2-35a the lower-level discriminator circuit consists of a Schmitt trigger (Q1 and Q2, see Sec. 1-19), a gate (an OR gate containing Q4 and Q5 with an emitter follower Q3), and a feedback amplifier (Q6 and Q7). The upper-level discriminator circuit consists of a Schmitt trigger (Q8 and Q9), a stretcher (Q10 and Q11), and a reset stage (Q12). Both discriminator circuits are connected to an anticoincidence circuit (Q13 and Q14). The analyzer generates an output signal when the lower-level trigger recovers. The lower gate and the feedback amplifier are combined into a feedback system to force the trigger to recover only at the zero input level, which is the condition for zero time shift. The waveforms in the analyzer are shown in Figure 2-36. The stretcher is a one-shot multivibrator which generates a rectangular output pulse when the input pulse firom the amplifier is over the upper level $E + \Delta E$. The reset stage terminates the rectangular pulse produced by the stretcher when the lower-level trigger recovers.

When the push button marked Diff. (Sw-1) is pressed, the analyzer is placed in its differential mode. The source of the window supplies a positive voltage to the base of $Q9$ in the upper discriminator. The cathode of diode $D3$ is grounded. The source of the threshold supplies a negative voltage to the bases of the NPN transistors Q8 and Q1 in the upper and lower discriminators, respectively. Then both Q8 and Q1 in the Schmitt triggers are normally cut off, and both Q9 and Q2 are normally conducting. The anticoincidence circuit consists of Q13 and Q14 connected in series, and Q13 is normally conducting while Q14 is cut off. When a

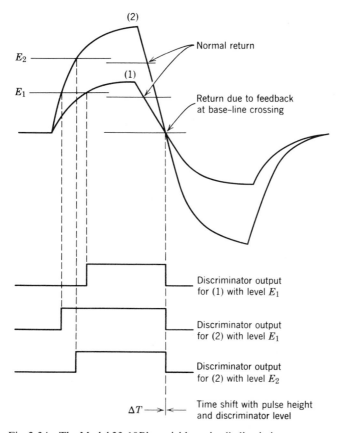

Fig. 2-34 The Model 33-10B's variable antiwalk discriminator outputs.

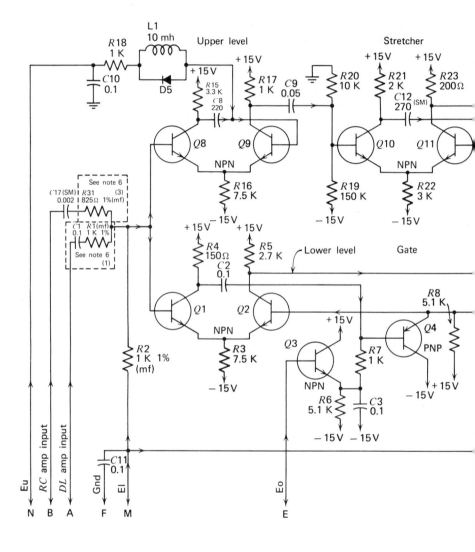

Fig. 2-35a Basic circuit of the RIDL Model 33-10B Antiwalk Single Channel Analyzer.

Notes
1. Transistor list.
   Q1,Q2,Q8,Q9---GA568,Q3,Q6---2N1306,Q4,Q5---2N1307,Q7---2N1143,Q10,Q11
   ---2N1308,Q12---2N835. Q13,Q14---2N779
2. Diodes.
   D1,D2,D4---1N916,D3---1N625,D5---RIDL979.

206

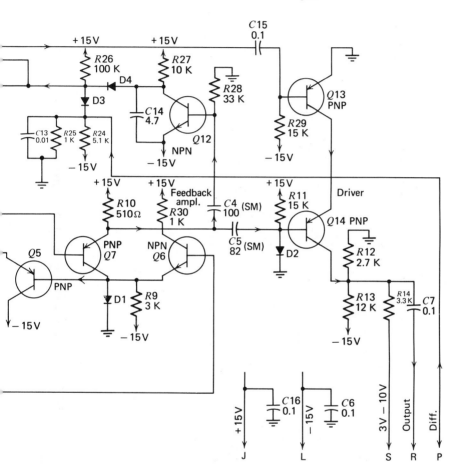

3. Last number used:
   C16, D4, R31, Q_{14}, L1
4. Numbers not used.
5. "SM" denotes silver mica cap. & "MF" denotes metal film res.
6. Assemblies.
   (1) 861181 (DL AMPL input not used) cap. C1 & res. R1 deleted.
   (2) 861182 (as shown)
   (3) 862216 (R31 value to be 1K$_\Omega$)

207

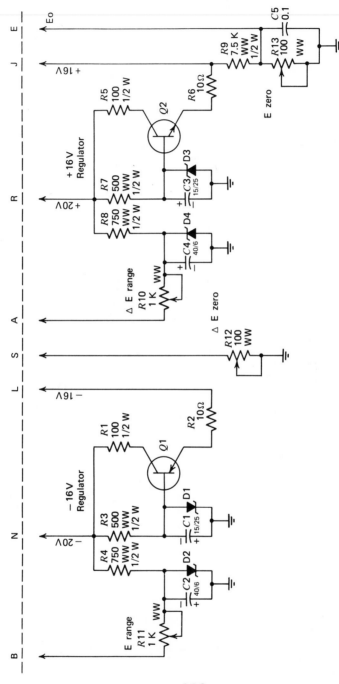

Fig. 2-35b   Regulator circuit of the Model 33-10B Single Channel Analyzer.

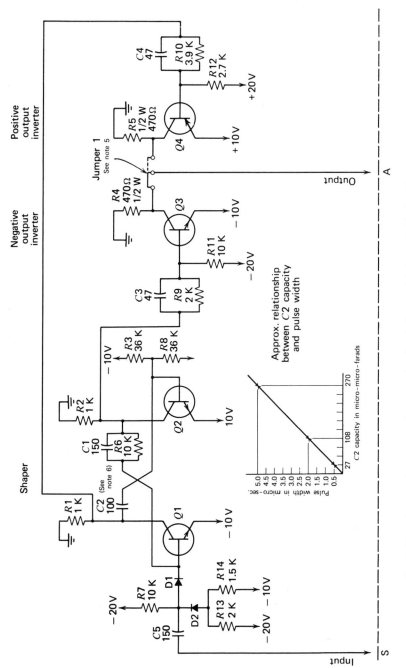

Fig. 2-35c Output shaper and inverter circuit of the Model 33-10B Single Channel Analyzer.

209

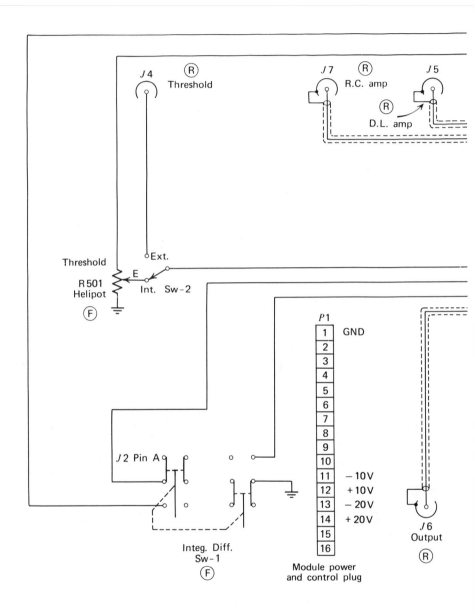

Fig. 2-35*d* Interconnection of the entire system for the RIDL Model 33-10B Single Channel Analyzer.

210

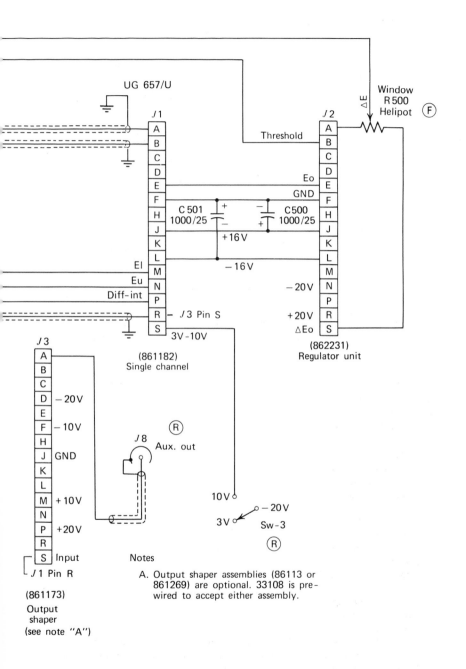

UG 657/U

J1

A
B
C
D
E
F
H
J
K
L
M — El
N — Eu
P — Diff-int
R — J3 Pin S
S

C 501
1000/25

C 500
1000/25

+16 V
−16 V

3V−10V

(861182)
Single channel

J2

A — Threshold
B
C
D
E — Eo
F — GND
H
J
K
L
M
N — −20 V
P
R — +20 V
S — ΔEo

Window
R 500
Helipot (F)

ΔE

(862231)
Regulator unit

J3

A
B
C
D — −20 V
E
F — −10 V
H
J — GND
K
L
M — +10 V
N
P — +20 V
R
S — Input

J8 (R)
Aux. out

10 V
3V

−20 V
Sw-3

(R)

J1 Pin R

(861173)
Output
shaper
(see note "A")

Notes

A. Output shaper assemblies (86113 or
861269) are optional. 33108 is pre-
wired to accept either assembly.

211

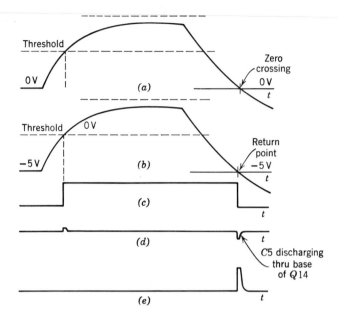

Fig. 2-36   Waveforms in the basic circuit of the Model 33-10B Single Channel Analyzer (Figure 2-35a): (a) input signal at point A or B; (b) signal at base of Q1; (c) signal at emitter of Q7; (d) signal at base of Q14; (e) output pulse from collector of Q14. Note that the base line is assumed to be at 5 volts (threshold potentiometer at 500 divisions).

Notes for Fig. 2-35b
1. Transistors
   Q1---2N1183,Q2---2N696
2. Diodes
   D1,D3---1N719A,D2,D4---1N752
3. Last numbers used
   Q2,C5,D4,R13
4. Numbers not used

Notes for Fig. 2-35c
1. Transistors
   Q1,Q2---2N1308,Q3---2N1304,Q4---2N404
2. Diodes
   D1,D2---1N90 or RIDL type 979
3. Last numbers used
   R14, C4, D2, Q4
4. Numbers not used
5. Assemblies
   (1) 861173   (negative output) jumper 1 connected as shown (solid line)
   (2) 861269   (positive output) jumper 1 connected as shown with dotted lines
6. For pulse width versus C2 capacity see chart

positive pulse which has an amplitude within a selected voltage window is applied to the bases of Q1 and Q8, it is not sufficient to trigger the upper Schmitt circuit, and thus no output pulse is produced by the upper discriminator. However, the lower discriminator is triggered, so that an output positive pulse is produced in the driver stage (Q14) by the trailing edge of the discriminator output (emitter of Q7).

When the input pulse has an amplitude above the upper level, the upper-level discriminator triggers, which in turn triggers the stretcher. The stretcher cuts off Q13, preventing the output to be generated.

PROBLEMS

2-1   Assume that a signal applied to the input stage of Figure 2-6a is the charge obtained from a cooled germanium detector ($\epsilon = 2.9$ eV) in a beam of 5-MeV particles and that the detected count rate is 10,000 cps. Find the rms fluctuation of the instantaneous baseline.

2-2   Figure 2-6b indicates a preamplifier with an emitter-follower output stage. It is shown that the rms value of input signal is 0.64 V and that the output impedance of the emitter-follower is $0.025/I$, where I is the collector current in amperes. With zero output signal and a "standing current" of 5 mA, the output impedance is 5 Ω. With a ±0.64-V base-line excursion, the current furnished to the external load is ±0.64 V/200 Ω = 3.2 mA. Find: (a) the range over which the current I_C swings; (b) the range of output impedance; (c) the percentage of variation in the heights of pulses arriving at the main amplifier; (d) the percentage by which those pulses contained within ±1 standard deviation (68 percent of the total pulses which occur) will be modulated; and (e) the percentage by which the remainder (32 percent) will be affected.

Ans. (a) 5 ± 3.2 mA: (b) 3 to 14 Ω;
 (c) 5.5 percent;   (d) 5.5 percent or less;
 (e) more than 5.5 percent.

2-3   Using the conditions of Prob. 2-1 in (2-8) and assuming an interstage coupling time constant of 50 μsec, find the rms voltage fluctuation for a second-stage gain of 2 and 16.

Ans. 0.29 V; 2.3 V

2-4   Try to get double integration by changing some component(s) in the circuit of Figure 2-28. Specify the value of the component(s) changed.

REFERENCES

Fairstein, E., and J. Hahn, Nuclear Pulse Amplifiers — Fundamentals and Design Practice, Parts I, II, III, *Nucleonics,* **23:** 7, 56 (July, 1965); 9, 81 (Sept., 1965); 11, 50 (Nov., 1965). Parts IV, V, *Nucleonics,* **24:** 1, 54 (Jan., 1966); 3, 68 (March, 1966).

Chase, R. L., *Nuclear Pulse Spectrometry,* McGraw-Hill, New York, 1961.

Price, W. J., *Nuclear Radiation Detection,* 2d ed., McGraw-Hill, New York, 1964.

Overman, R. T., Laboratory Manual, *Semiconductor Detectors and Associated Electronics Introduction to Theory and Basic Applications,* ORTEC Incorporated, Oak Ridge, Tennessee.

Radeka, V., Low-Noise Preamplifiers for Nuclear Detectors, *Nucleonics,* **23:** 7 (July, 1965).

Littauer, R., *Pulse Electronics,* McGraw-Hill, New York, 1965.

Nowlin, C. H., and J. L. Blankenship, *Rev. Sci. Instr.* **36,** No. 1830 (1965).

Sherman, I. S., ANL, A Versatile Preamplifier for Semiconductor Detectors Providing High Resolution Capacity at High and Low Counting Rates, EL-7-14.

Strauss, M. G., ANL, Solid-State Pulse-Height Encoding System with Pileup Reduction for Counting at High Input Rates, *Rev. Sci. Instr.,* **34:** 4, pp. 335–345 (April, 1963).

Van Valkenburg, M. E., *Network Analysis,* Prentice-Hall, Englewood Cliffs, N.J., 1962.

Instruction Manual, *Model 33-10B Anti-Walk Single Channel Analyzer,* FM-1106, Radiation Instrument Development Laboratory, 1963.

CHAPTER **3**

MULTICHANNEL
PULSE-HEIGHT ANALYZERS

3-1 MULTICHANNEL PULSE HEIGHT ANALYSIS

Introduction

Nuclear pulse spectrometry may be carried out using a single-channel or a multichannel pulse-height analyzer. The single-channel analyzer (SCA) is a very useful tool for routine counting and for scanning samples to determine the energy spectra of radioactive sources. About 15 to 30 min are required per scan. This is because the SCA is capable of recording only those pulses falling within a single channel while all other pulses are rejected. Thus to cover 100 channels, one must count point by point 100 individual channels. This takes time. Therefore when energy spectra of low intensity are studied, inordinately long counting times are required to obtain a certain accuracy. This time element makes the use of a SCA impractical for low-activity samples, as well as for specimens with short half-lives.

Multichannel analyzers (MCA's) have been developed which help to avoid this limitation. In a MCA determinations of the counting rates can be made simultaneously in all channels; that is, they are capable of recording virtually every pulse from an energy spectrum as the pulses occur. The only pulses not recorded are those which occur while the analyzer is busy handling a previously acquired pulse. Since such analyzers can be constructed to require analysis time on the order of 10 μsec/pulse, few pulses are ignored when reasonable source intensities are employed (up to about 30,000 pulses/sec).

The analyzers usually have 400 or 512 channels for single-parameter devices and thus time can be saved when all channels are permitted to accumulate information. The result of a single-parameter multichannel analysis is a spectrum, usually in the form of a histogram, with number of events given as a function of one or more of such parameters as the height of the pulse, its shape, its time of occurrence, or of the height, shape, or time of occurrence of coincident or otherwise time-related events. Multiparameter multichannel analysis has for some time been used by the larger laboratories. In determining a complex nuclear decay scheme, one can examine the source with two detectors whose outputs serve as the two inputs to a two-parameter analyzer. When the events involving coincident pulses in the two counters are stored, there results a three-dimensional record from which can be extracted a great deal of information regarding the intercorrelation of the events in the two detectors.

All single-parameter multichannel analyzers consist of some or all of the following elements: (a) an analog (pulse-height)-to-digital converter which associates the height of each input pulse with a specific channel; (b) a memory- or data-storage device which preserves the number of pulses which fall in their associated channels and, thus, contains information on the number of pulses in each channel until this information is recorded otherwise; (c) a data-display device which gives the experimenter information about the data stored in the memory without removing it from the memory; (d) auxiliary data readout devices such as curve (X, Y) plotters, paper-tape or card punches, magnetic tape, typewriter, etc.

Basic Principles of Operation

The first multichannel analyzers employed multidiscriminator converters, a logical extension of single-channel analyzers. By employing n differential discriminators set at successive window levels and each having its own counting system, an n-channel analyzer is obtained. However, such an arrangement is costly to construct and to maintain. Also, it is a difficult design to keep in adjustment, particularly from the standpoint of maintaining equal window widths. Figure 3-1 shows the chief components of a modern one-parameter multichannel analyzer.

The most widely used system employs the pulse-height-to-time con-

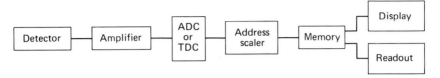

Fig. 3-1 One-parameter multichannel analyzer. ADC and TDC stand for analog-to-digital converter and time-to-digital converter, respectively. TDC is used in time-of-flight work.

verter as the analog-to-digital converter (ADC). This converter was developed by Wilkinson and Hutchinson and Scarrott. Figure 3-2 illustrates the process of the Wilkinson type of analog-to-digital conversion. First, a small capacitor is charged up to the peak voltage of the pulse; it is then discharged at constant current. While the discharge is in progress, clock pulses from a stable oscillator are counted by a scaler, the number of clock pulses counted being proportional to the time the capacitor took to discharge and hence to the original height of pulse.

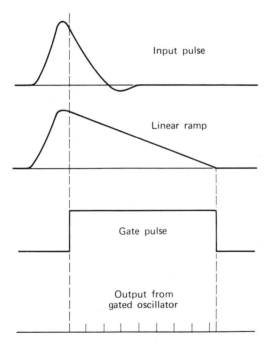

Fig. 3-2 Illustrating the discharge type of analog- (pulse-height-) to-digital converter (ADC).

Another way of looking at this is that a capacitor is charged linearly until it reaches the height of the stretched input pulse. By starting the ramp in synchronism with a clock pulse, a one-channel uncertainty in the start time can be avoided.

Thus the signal pulse shown in Figure 3-3 is converted to a flattopped one, *b*, by means of a pulse-stretching circuit. At the start of the pulse, a linearly rising signal *c* is started, and a series of equally spaced pulses *d* is switched on. When *c* reaches *b*, the pulse train *d* is stopped. In this manner, the pulse heights are converted into trains with different numbers of clock pulses. These trains of pulses can be sorted, stored, displayed, and read out by various techniques that have been developed for digital computers. The number of clock pulses that occurred before the oscillator gate closed is the number of the channel that the count is to be stored in. This number is called the address.

Modern day one-parameter analyzers tend to use memories consisting of three-dimensional arrays of small ferromagnetic cores. Typically, the memory cores of one address, as selected by the address scaler (register), are read out during the read current pulse when the information from the cores is transferred to the memory register. Thus a count is stored by reading out the contents of the appropriate channel, adding one, and restoring the result. This interrogation, the restorage process, takes perhaps 10 to 30 μsec. But a much greater contribution to pulse-height-analyzer dead time is the result of the conversion process. The faster clock oscillator circuits in general use give pulses at a 4- or 5-Mc rate, making the conversion time for a pulse in channel 400, for instance, equal to 80 or 100 μsec. Thus it is necessary to prevent new pulses from entering the converter while it is converting, so that storage rates are limited to less than 10^4/sec.

In time-of-flight work the type of time-to-digital converter (TDC) used depends to some extent on the magnitude of the flight times involved. For flight times running, say, 20 μsec or longer, it is feasible to start a clock oscillator at time zero (when the neutron starts its flight) and to stop it when the neutron is detected. The count is then stored as described before. For shorter flight times, ranging down into the nanosecond region, digital circuits are not fast enough. However, one can start a linear ramp at time zero and stop it when the neutron arrives. The result is a pulse whose amplitude is proportional to the time of flight, and it can be analyzed by conventional pulse-height analysis methods.

To convert the digital information of the address and memory registers from digital into analog form, two digital-to-analog converters (DAC's) are usually provided. These are the horizontal and vertical DAC's, re-

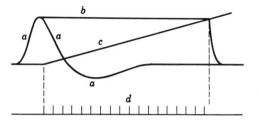

Fig. 3-3 Typical waveforms in sweep type of analog-to-digital converter. (From R. L. Chase, *Nuclear Pulse Spectrometry*, p. 82, McGraw-Hill, New York, 1961.)

ferred to as the address decoder and data decoder, respectively. A typical DAC is shown in Figure 3-4. Suppose that, say, a train of 23 clock pulses is applied to the input of the counting chain. In binary notation this number is written 10111. Accordingly, $B4$ is in state 1, $B3$ in state 0, $B2$ in state 1, etc. Assume that when a binary is in state 1, its Y output is higher by ΔV than when it is in state 0. Since at the input of the operational amplifier we see a virtual ground (Sec. 1-14), the change

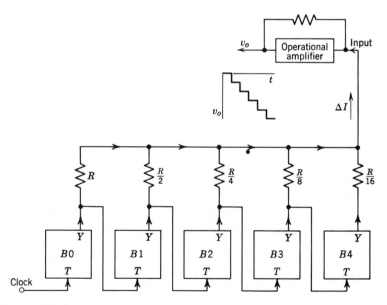

Fig. 3-4 A digital-to-analog converter. A voltage is generated which is proportional to the number of pulses in a train.

in amplifier input current ΔI over that corresponding to the situation when all binaries are in state 0 is

$$\Delta I = \frac{\Delta V}{R/16} + \frac{\Delta V}{R/4} + \frac{\Delta V}{R/2} + \frac{\Delta V}{R}$$

$$= (16 + 4 + 2 + 1)\frac{\Delta V}{R} = 23\,\frac{\Delta V}{R}. \qquad (3\text{-}1)$$

Thus the input current change to the amplifier and its output voltage v_0 are proportional to the count registered in the scaler. As the input pulses are applied, the amplifier output is the stair-step voltage indicated. The output will jump a fixed increment each time the count increased by unity.

Using the principle given above, the position of the address and memory registers is generally converted into a current, linearly related to the position of these registers: the higher the channel number or the amount of counts, the higher the current. The current is then converted into a voltage by the operational amplifier.

Features of One-Parameter Analyzers

In one-parameter analyzers the number of channels ranges from 200 to 800, with capacities of 10^5 or 10^6 counts per channel. Most instruments offer selective storage—the ability to divide the memory into halves, quarters, or eighths, and to "route" the input to any desired segment. Thus it is possible to store up to eight spectra in the same memory. Background subtraction can be accomplished by first storing background in the negative mode and then counting the sample for the same length of time in the positive mode. The feature of memory transfer, combined with selective storage, permits one to subtract the same background count over and over, by storing it in one segment of the memory, in the negative mode, and then transferring it nondestructively (i.e., without erasing the original record) to the segment in which one will store the spectrum of the sample. Automatic programming may also be carried out when counting a series of samples using an automatic sample changer.

Spectrum stripping is the process of analyzing a spectrum by subtracting from it, one by one, the spectra of known components. Some systems permit this to be done as follows. First the standard spectra of the components to be looked for must be individually stored on paper tape. Then the spectrum to be analyzed is stored in one-half of the memory. The standard spectrum of component A is then read from the tape into the other half of the memory. A computer in the analyzer permits any desired multiple of the standard spectrum to be subtracted from the unknown. The operator observes the process on the oscilloscope display, and subtracts in steps until the features of component A seem to have been cancelled. The process can then be repeated for the next component. An important point to be aware of here is that, while ordinate (intensity) multiplication of the standard spectrum is provided, abscissa (pulse-height) multiplication is not. It is, therefore, imperative that the standard and unknown spectra have the same resolution, with the same peaks occurring at the same locations. A difference of even 0.5 percent in peak position can cause unsatisfactory results.

Most of these analyzers have two or more inputs, with internal routing, making it possible to store several spectra simultaneously (but not simultaneous pulses, of which at least one must usually be rejected), with each input routed to its own memory segment. In practice, it is difficult to eliminate completely cross-talk storage of the occasional pulse in the wrong memory segment.

All the analyzers can display the memory contents on a cathode-ray tube (CRT) while no input is being analyzed. In addition, most of them have a live display, which shows each pulse as a brief flash on the cathode-ray screen as it is being stored. If the counting rate is fairly high, one can watch the spectrum growing.

Typically a switch is provided to render the ordinate (relative intensity scale) either linear or logarithmic. The latter has the advantage that the displayed spectral shape is independent of the counting time. This is particularly useful in visually comparing a spectrum stored in one segment of the memory with a similar one in another segment, thus utilizing a feature which permits the simultaneous, superimposed display of more than one segment.

While the analyzer is busy converting and storing a pulse, it is rendered insensitive to new input pulses. For accurate quantitative work it is often insufficient to know the clock-time duration of the count: one needs to know the live time, which is the clock time minus the dead time—the actual length of time the analyzer was ready to accept input pulses. This can be measured rather simply by means of an accurate, usually crystal-controlled, free-running oscillator, whose pulses are counted only if the input is "live" when the pulse occurs. Thus, within statistics, the number of live-timer pulses accepted is proportional to the live time, provided that the pulses to be analyzed arrive at random times. If the pulses arrive in bursts, as, for instance, from a pulsed accelerator, the live timer will give a misleading indication.

In most analyzers, a meter continuously indicating present dead time is either provided or can easily be added. Some instruments print out the live time of the run as the contents of the first data channel and some do not.

The option of multichannel scaling is offered by most of the instruments. With this feature in operation, the channels become time channels instead of pulse-height channels. At the start of counting, all input pulses above the lower discriminator setting are registered as counts in channel 1. Channel-advance signals, internally or externally generated, switch the counting to channel 2, and so on, at a predetermined rate. The result is a record of the counting rate as a function of time such as one would want for determining the half-life of a radioactive source.

For quantitative analysis, the area under a peak is often a more reliable indication of its intensity than the amplitude at the maximum. Some instruments can be programmed to integrate under peaks, summing the counts between selected pulse-height limits.

All types of readout equipment are available: oscilloscope camera, punched paper tape, typewriter, parallel paper-tape printer, magnetic tape, *XY* plotter. The punched-tape and magnetic-tape readouts can be made compatible with the input required by computers.

In using the cathode-ray-tube display or *XY* plotter, one should be aware that some analyzers "decode" for display the two most significant digits only, while the others decode three. With two-digit decoding, there is the possibility of a maximum error of 10 percent in the ordinate for any one channel, since 109 counts appears as 100 counts, 1099 as 1000, and so on.

Performance of Single-Parameter Multichannel Analyzers

The numbers given below are indicative only, since there are differences between instruments.

1. Integral linearity: $\pm$ 0.5 percent of full scale, over top 97 percent of scale. Integral linearity is the maximum deviation of a channel position from where it should be as a percentage of full scale.

2. Differential linearity: $\pm$ 2 percent, over top 97 percent of scale. Differential linearity is the maximum deviation of channel width from the average for all channels. Channel-width variations can cause spurious structure to appear in the spectrum.

3. Dead time: $(30 + 0.5 \text{ N})$ μsec. From this, one can tell that in this particular case the ADC clock pulses come at a 2-megacycle rate, and that 30 μsec are required to store the count and allow the circuitry to settle down in preparation for the next pulse. N is the number of the channel that the count is to be stored in.

4. Zero drift: less than 0.1 percent of full scale per 24 hr, and less than 0.4 percent of full scale per 10°F. The zero drift is the shift of the entire spectrum left or right.

5. System gain drift: less than 0.1 percent of full scale per 24 hr, and less than 0.4 percent of full scale per 10°F.

6. Maximum input rate: up to 50,000 counts per second, "with negligible spectrum distortion." Nevertheless, for quantitative work with present-day circuitry, the input rate should usually be kept well below 10^4 counts per second.

7. Input-pulse-shape requirements: variable from one instrument to another, but in general it can be said that pulses of the same height but with very different rise or fall times will tend not to count in the same channel. Many instruments do not function in top form if the pulse rise time is longer than a microsecond.

8. Live timer accuracy: better than 0.5 percent, up to 5000 counts per second.

9. Multiple-input cross-talk rejection: some instruments are better than others.

10. Power consumption: less than 50 watts.

In many cases, the numbers given above represent specifications set by the manufacturers. Some fast analyzers (such as HP Model 5400A) have been developed with better performance (see Table 3-1).

Multiparameter Analyzers

The advent of multiparameter analyzers has greatly speeded data taking in certain types of experiments, and has, in fact, made possible a new range of experiments that could probably not be done otherwise. For the most part, multiparameter analyzers use the principles of operation developed for single-parameter instruments, the chief difference being in the required memory size. The magnitude of the difference can be appreciated by noting that for 100-channel coverage in each of n dimensions, one needs 100^n channels, making 10^4 channels for two dimensions and 10^6 for three. Frequently 100-channel resolution is not good enough.

Memories for multichannel analyzers can be of two types—accumulating and open-ended. As examples of the former we have the commonly used magnetic-core memory and the less widely used rotating magnetic drum. In such a memory, each channel has its own set of locations, and whenever a count occurs, the contents of the appropriate channel are increased by one. A typical open-ended system records all the information about each event digitally on magnetic tape (or occasionally on punched paper tape, moving film, etc.). An accumulating memory has the distinct advantage that its contents can be interrogated at will, to see how the experiment is going, whereas the open-ended tape must be processed later in auxiliary equipment.

Figure 3-5 shows an experimental arrangement for two-parameter analysis.

As an example of one of the more elaborate arrangements, the Argonne

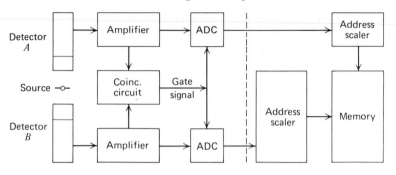

Fig. 3-5 Typical experimental setup with a two-parameter analyzer.

National Laboratory three-parameter analyzer will be briefly described. This instrument achieves more than 33 million (2^{25}) channels by storing each event on a 25-track magnetic tape that is $1\frac{1}{4}$ inches wide. For one of the parameters 512 channels are available, leaving 256 for each of the other two. After the storage of each event the tape is advanced by 0.01 in. to be ready for the next one. A one-event buffer storage gives a two-event resolving time of 230 μsec, which is determined largely by the time required for analog-to-digital conversion. At high counting rates the tape moves continuously. Maximum storage rate is 170/sec, which, because of the usual slow rate of detection of three-parameter events, is satisfactory.

The system was designed explicitly for use in studies of the gamma-ray spectra that result from the capture of neutrons in resonances. A neutron beam from a reactor is divided into bursts by a rapidly rotating "chopper." Twenty-five meters from the chopper a sheet of the material under investigation intercepts the beam, and when a neutron is absorbed, the resulting excited nucleus usually gets rid of its excess energy immediately by emitting several gamma rays. Two large sodium-iodide scintillation crystals detect some of these gamma rays. When a coincidence occurs, its time of occurrence is recorded on the tape (using the time-to-digital conversion process previously described) as one of the three parameters, the other two being the pulse heights in the two detectors. The time of occurrence gives the energy of the neutron which caused the event, and this can be correlated with information about the nuclear structure as revealed by the two pulse heights.

After the data have been accumulated (typically in one or two days of running time) the tape is transferred to a "search station," which might be described as a special-purpose digital computer. The search station permits four two-dimensional (that is, one-parameter) spectra to be read

simultaneously from the tape into a 1096-channel core memory in about
$3\frac{1}{2}$ minutes, and the memory contents can be displayed on an oscilloscope or read out for a permanent record. Each such spectrum is the result of imposing operator-selected conditions on two of the three parameters — say a particular neutron time of flight and a selected pulse-height
interval in one of the detectors — with the recording of the entire spectrum
in the third parameter of events which met the two imposed criteria.

3-2. INTRODUCTION TO THE TYPICAL ONE-PARAMETER ANALYZERS

RCL Model 20631/2 400-Channel Analyzers

GENERAL. The basic analyzer includes the following units: (a) a
linear pulse amplifier; (b) an analog-to-digital converter; (c) a ferrite
core memory; (d) an address register; (e) a data or memory register;
(f) the digital-to-analog converter; (g) the control logic; (h) the read
logic; (i) the timer. The linear amplifier amplifies pulses from a low-level
signal source to the 5-volt level required by the ADC. The ADC converts a signal pulse into a pulse train. The number of pulses in the pulse
train is linearly proportional to the signal pulse amplitude. One count is
added to the ferrite core memory in the channel number corresponding to
the number of pulses in the pulse train. The address register counts the
pulses from the ADC pulse train and controls the memory. The data or
memory register is used as a temporary storage for the channel (selected
by address register) in which the data must be changed (read out or read
in). The digital-to-analog converters convert the address-register and
memory-register data into an analog signal for display on an oscilloscope
or for readout on a recorder. The control logic allows the operator to
perform the following functions: (a) accumulation of data in the memory;
(b) displaying of the accumulated data; (c) reading out the accumulated
data; (d) automation of the ACCUMULATE-READ cycle. The built-in
timer with its preset controls can be used to automatically terminate the
accumulate mode of operation.

THE ADC. The Model 21106 ADC is compatible with the Model
20631/2 analyzers. It is shown in Figure 3-6 as a simplified block diagram. The single-channel input is usually used for coincidence analysis.
The coincidence input control and the upper- and lower-level discriminators form the single-channel circuit. The input to ADC control chooses
the signal source. The internal amplifier accepts pulses from a scintillation
detector with a preamplifier, or from a similar source. For coincidence
operation a gate signal is required for analysis. The zero intercept control allows zero energy to be placed in channel zero. One output of the

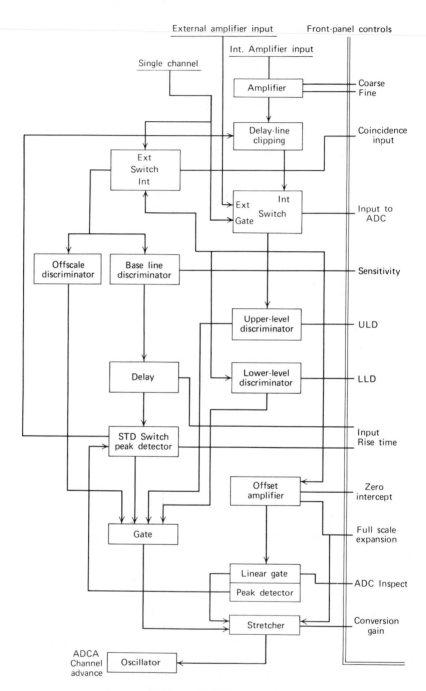

External amplifier input   Front-panel controls

Int. Amplifier input

Single channel

Amplifier — Coarse / Fine

Delay-line clipping — Coincidence input

Ext Switch Int

Ext / Gate — Int Switch — Input to ADC

Offscale discriminator

Base line discriminator — Sensitivity

Upper-level discriminator — ULD

Delay

Lower-level discriminator — LLD

STD Switch peak detector — Input Rise time

Offset amplifier — Zero intercept

Full scale expansion

Gate

Linear gate — ADC Inspect

Peak detector

Stretcher — Conversion gain

ADCA Channel advance — Oscillator

Fig. 3-6   Block Diagram of Model 21106 ADC.

226

offset amplifier drives the two signal discriminators (off-scale and base-line). The conversion gain control changes the stretcher capacitor. A stable 2-Mc (MHz) oscillation is generated by a constantly running oscillator. This oscillation is broken up into a series of equal length pulses to form a pulse train. The number of pulses in the pulse train is proportional to the height of the incoming pulse. The resulting pulse train is the digital output of the ADC that goes to the analyzer. This mode of operation is the pulse-height analysis mode of the analyzer. The Model 21106 ADC can also be operated as a multichannel scaler.

ADDRESS REGISTER. The address register determines which address in the memory is interrogated. This register counts the pulses in the pulse train of the ADC (ADCA) or the pulses from the display clock. The 400 addresses are selected by two-decade scalers puls two bistables, referred to as units register, tens register, and 100 and 200 bistables. The basic flip-flop stages in the scalers are either similar to, or the same as, that shown in Figure 1-126. For every 10 counts received by the units register, the tens register will receive one count. After 10 counts from the tens register, the 100 bistable will receive a trigger and so will the 200 bistable after two counts from the 100 bistable.

When the tens register has received 10 carry pulses from the units register, the 100 bistable will be switched to the "1" state. The units and tens registers are at the "0" state. When 200 pulses have been counted, the 100 bistable is reset to "0" and the 200 bistable is set to the "1" state. In this way 399 pulses can be counted.

The address overflow signal is used to stop the multiscaler mode and the readout mode at the end of the selected group of channels. In the AUTO mode this same signal will advance or recycle the program during multiscaler or readout. It is also used as a strobe signal for the GROUP switch.

THE MEMORY. The memory is the digital storage unit for the analyzer and is made up of 400 channels, addresses, or words. Every address consists of 20 or 24 bits, making up a storage capacity of $10^5 - 1$ or $10^6 - 1$ counts per address. The basic element of the memory is a ferrite core, as shown in Figure 3-7a. This core can be magnetized in two directions by means of a current in a wire passing through the core. After a core has been magnetized, no current is required to maintain this magnetism. Reversing the current will magnetize the core in the other direction. These two stable states therefore make the ferrite core a typical storage device. The cores are assembled into 20 or 24 arrays of 400 cores, in which the 400 cores are arranged in squares of 20×20. Each core is one bit of a different address. The total amount of cores for a 400-address memory and a storage of $10^6 - 1$ counts is 400×24 or 9600.

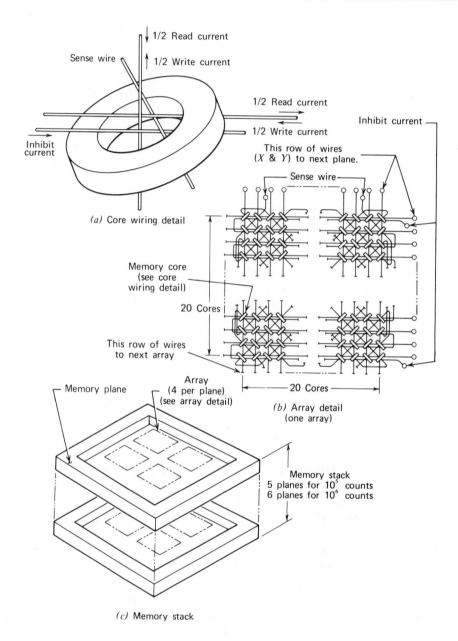

(a) Core wiring detail

1/2 Read current

Sense wire

1/2 Write current

1/2 Read current

Inhibit current

1/2 Write current

Inhibit current

This row of wires
(X & Y) to next plane.

Sense wire

Memory core
(see core
wiring detail)

20 Cores

This row of wires
to next array

Array
(4 per plane)
(see array detail)

20 Cores

(b) Array detail
(one array)

Memory plane

Memory stack
5 planes for 10^5 counts
6 planes for 10^6 counts

(c) Memory stack

Fig. 3-7   Mechanical construction of the memory.

In order to select one of these 400 cores, two wires (the X and the Y wire) are passing through every core at a 90° angle. There are 20 X Wires and 20 Y Wires, with a ferrite core at the crossing point of each X and Y wire. The selection of these two wires is done by the address register. By extending the X and Y wires through all the 20 or 24 arrays, it is possible to select 20 or 24 cores at the same time in one particular address. If the selected X and Y wires are each supplied with half of the necessary magnetizing current, all the selected cores will switch in one direction. Only one X and one Y wire are selected at a time.

Figure 3-8 shows the memory drivers, which are bilateral transistors ($Q2$ and $Q3$) that drive the XY lines of the memory and which are controlled by the BCD-to-ten line decoder of the address register. The bilateral transistor is capable of carrying current in either direction.

Every 20×20 array in the memory is divided into quarters, as shown in Figure 3-7b. The quarters in which the selection of XY wires occurs depends on the selection of the 100 and 200 bistables of the address register, each driving two bilaterals. Depending on the above, the selected core will be in the 0 to 99, 100 to 199, 200 to 299, or 300 to 399 section of the memory.

At the "read" time, the selected memory wires are supplied with a positive (as seen from the power supply) current from a constant-current generator controlled by the read pulse. The selected bilaterals are supplied with a negative collector current during the "write" time.

During a read cycle any of the selected cores switched from the 1 state into the 0 state will induce a current pulse in the sense wire for that array of 400 cores. One sense wire is going through every core in one array. There are 20 to 24 sense wires. The cores that have only one of the selected address wires passing through will not be switched because half of the read current is not sufficient.

As the sense wire passes through the cores in alternate directions to cancel false signals caused by half-selected cores, the signal on the sense wire is bipolar. A sense wire is connected to the input of a sense amplifier. There are 20 or 24 sense amplifiers which are all differential amplifiers for the rejection of common mode signals. The two outputs are combined in an "OR" gate to divide one polarity pulse. The output is gated with the MR READ STROBE to eliminate pulses on the sense wires which are not caused by the complete switching of a core during the read time. Every sense amplifier will set one bistable (one bit) of the memory register (see Figure 3-19).

During a write cycle the current flows in the opposite direction, and this will switch all the cores to the "1" state unless commanded by the analyzer to leave the core in the "0" state. This command is given by a

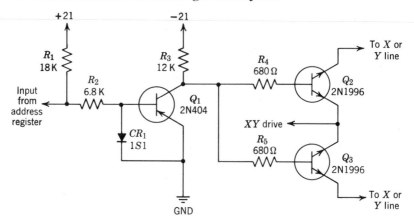

Fig. 3-8   Typical memory driver circuit. Transistors $Q1$ (inverter), $Q2$, and $Q3$ (bilaterals) comprise a typical memory driver circuit. To drive a 400-channel memory it would require 20 of these circuits, 10 for the X drivers, and 10 for the Y drivers.

current pulse through the inhibit wire, which is in coincidence with the write pulse. The inhibit wire is passing through all the cores of an array in such a way that it opposes the current in one of the two address wires. The resulting current is inadequate to switch the core to the "1" state. Thus a "0" is written with inhibit current, and a "1" is written without inhibit current. The inhibit wires are commanded by the memory register. The mechanical construction of the memory is as in Figure 3-7. Every array has one sense wire and one inhibit wire.

MEMORY REGISTER.   This register is used as a temporary storage for the ferrite core memory. Information is counted or transferred into the register. The memory cores of one address, as selected by the address register, are read out during the read current pulse when the information from the cores is transferred to memory register. During the write portion of the memory cycle the information from the memory register is written back into the memory by means of the inhibit drivers. The memory register is a five- or six-decade scaler with add and subtract logic built in.

TII Models 402 and 404 400-channel Analyzers

Figure 3-9 shows the front panel of the TII Model 404C analyzer.

INPUTS.   The 402C and 404C analyzers accept either negative pulses between 0 and 50 mV at maximum gain or positive pulses between 0 and 10 V or 0 to 100 V with a special attenuator. Selection of the appropriate

Selects memory subgroups
1/1 = 0-399
1/2 = 0-199
2/2 = 200-399
1/4 = 0-99
2/4 = 100-199
3/4 = 200-299
4/4 = 300-399
A/2 = Automatic routing to subgroups
1/2 from BNC 1 or 2 or 2/2
from BNC 3 or 4 on "first come,
first serve basis."
A/4 = Automatic routing to subgroups
1/4 from BNC 1
2/4 from BNC 2, etc. —— S2

IMPORTANT—Selection of any memory group sustains over-all energy calibration, e.g., if instrument is calibrated 0-2MEV with 400 channels in use (1/1), then switched to a subgroup of 200 channels the calibration would remain exactly 0-2MEV; only the number of energy increments would change. Likewise for any 100 channel subgroup.

Connects internal pulse generator directly —— S1
to input connectors.

Restricts maximum input signal to be —— R6
analyzed.

Sets channel position of maximum signal —— R1
to be analyzed.   R3 R2 R4

Sets channel position of minimum input —— R5
signal to be analyzed.

Four input connectors. Pulses from any input connector can be routed to any desired memory location such as 1/1, 1/2, 2/2, 1/4, 2/4, 3/4, or 4/4, by setting the memory location switch to the appropriate position. However, for automatic routing of multiple inputs, one precaution should be noted i.e., in the automatic half mode (A/2 Model) one connector each side of the center line must be used, i.e., 1 and 3, 2 and 3, 2 and 4 etc.

BNC1
BNC2
BNC3
BNC4

[1 — Flashes when input pulse rate exceeding upper level is greater than 250 pps.
R7 — CRT Controls
R8
R9 — Positions traces horizontally.
R10 — Expands horizontal trace approx. 4 times.
S3 — Selects linear or logarithmic display and determines the range of the linear display. The calibrate position introduces reference lines for the logarithmic display.
R11 — Determines vertical centering of subgroups
R12 of memory and operates in conjunction with COMPARISON switch.
S6 — For comparing spectra in various halves or quarters.
S5 — Selects Dwell Time Modes (1, 10, 100, 1,000 ms) of operation or EXT position for input of other Dwell Time periods or a PHA Mode.
S4 — Selects total "live time" accumulation.
S7 — Add-Subtract Live time.
S8 — Data Add-Subtract
S10 — Selects mode of Readout
S9 — Tests computer circuits by adding one count to each channel during each display cycle.
S13 — Resets memory during display
S14 — Resets memory during readout
S12 — Enables transfer of data during readout
S11 — Transfers data during display
SL4 — Stops all computer operations (does not turn off power).
SL3 — Starts readout

S15 — Is normally in the manual position. Upon initiation of an accumulate cycle, the switch is moved into the "Auto" position. At the end of the "Accumulate" period, the Analyzer will go from "Accumulate" to "Readout". At the end of the Readout period, the Analyzer will go into the Stop mode if S14 is in the "Down" position. The Analyzer will go back into "Accumulate" if S14 is in the "Up" position.

SL1 — Starts accumulation

SL2 — CRT display of information in memory

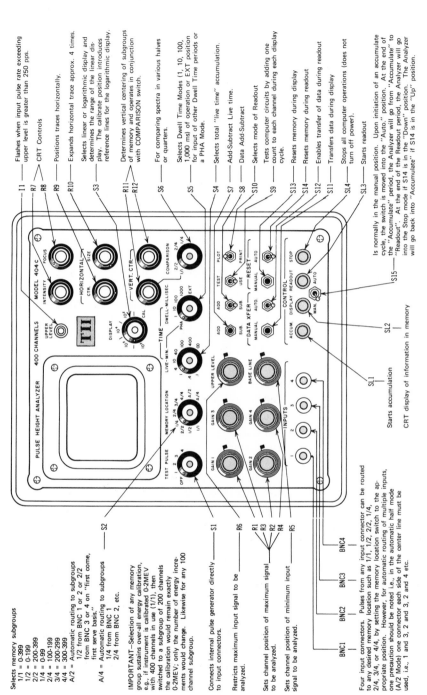

Fig. 3-9   Front panel of the TII Model 404C 400-channel analyzer.

231

cards[1] therefore depends on the amplitude and polarity of the input pulse. Low-level cards will accommodate negative input pulses ranging from 0 to 50 mV at maximum gain settings. Pulses above this potential will overflow the analyzer's sorting capacity and be lost. Accordingly, negative pulses greater than 50 mV must be attenuated before being directed into the analyzer. High-level inverter cards accept positive inputs between 0 and 10 V. Pulses from 0 to 100 V in potential require a 3.3 K inline attenuator to reduce the input signal to acceptable analysis levels. Both high- and low-level cards have a direct coupled impedance of 500 Ω. Detectors must be coupled to the inputs by a capacitor to prevent dc levels from influencing the bias on input circuits of either card.

PULSE SHAPES. The low-level amplifiers and high-level inverters have been designed to accept exponentially rising and decaying pulses. Ideally the pulse should have about a 0.5-μsec rise time (10 to 90 percent). The analyzer will accept and analyze other pulses which depart somewhat from the ideal, but the departure must not be too radical. The total width of the pulse presented should not be less than about 1.0 μsec at 90-percent amplitude points. By correct adjustment of the clip timing circuit, the analyzer can accept pulses for rise times up to 3 μsec. The analyzer can accept pulses with a decay time up to 20 μsec. However pulse "pileup" within the amplifier may result in causing base-line distortion at high counting rates. Flattopped pulses may be introduced, provided they meet both of the above conditions. Pulses with very fast rise times (less than 0.1 μsec) will tend to cause "ringing" or spurious oscillations and should be avoided.

MEMORY. The 400-channel analyzers use magnetic ferrite core memory systems which store the data in parallel binary-coded-decimal form. Information stored in this manner can be read out directly from the memory with a single "memory cycle" into a register connected to a printing device. Memory packages are made up of 24 planes or $10^6 - 1$ counts per channel with 400 cores mounted on each plane.

MEMORY GROUPING AND DATA TRANSFER. The memory of the analyzer may be split into four subgroups of 100 channels each, two subgroups of 200 channels each, or the entire memory may be utilized. If one input is used, it can be directed to any quarter of the memory, half of the memory, or the whole memory. If two inputs are used, each input can be directed to either a quarter or a half of the memory. If four inputs are used, each input may be directed to a respective quarter subgroup, that

[1]The fundamental logic circuits, which are used over and over again, are mounted on a number of plug-in units called logic cards.

is, input 1 to 1/4, input 2 to 2/4, and so on. Data which has been stored in one-half of the memory may be transferred manually to the other half. Data from one quarter may be transferred to the other quarter within that half.

ADD-SUBTRACT LOGIC. Memory information can be stored in either an add or subtract mode. A common application of this feature allows a standard or reference spectrum to be subtracted from a complex, unknown spectrum. The reference spectrum is accumulated in one subgroup in the "subtract" mode.

Hewlett-Packard Model 5400A Multichannel Analyzer

The Hewlett-Packard 5400A multichannel analyzer is a fast, accurate, versatile tool for spectrum-analysis work. It has a pulse-height analyzer (PHA) mode of operation, a sampled voltage analysis (SVA) mode of operation (for probability function analysis), and a multichannel scaling (MCS) mode of operation (for decay studies and Mössbauer work).

Built in modular form, it consists of an analog-to-digital converter plug-in housed in a power supply/interface mainframe, a digital processor, and a 50-MHz oscilloscope main frame and display plug-in. This modular "three-box" approach allows complete separation of the three main functions—digitizing the input, manipulation and control of the digital data, and visual display. This separation, logical grouping of function controls, and an annunciator function keying system permit operation of the front panel controls to be self-explanatory.

The HP Model 5415A analog-to-digital converter (ADC) is compatible with the 5400A analyzer. This ADC utilizes a ramp-type conversion technique and digitizing clock rate of 100 MHz. It will accept positive pulses with a full scale range of 1.25 to 10 V. The full-scale output range from the ADC is from 128 to 1024 channels. Upper- and lower-level discriminators are provided for simple discrimination against noise or other undesired inputs. The 5415A has a leading edge variable time-to-peak control to match the input pulse shape to the ADC. Base-line offset controls are provided to allow further flexibility in matching the accumulated data to a desired format. These controls allow continuous offset of the base line from −1.25 V to +10 V. The ADC front panel is shown in Figure 3-10 (left).

The HP Model 5410A power supply interface is designed to contain the 5415A ADC, to provide power to the digital processor, and to contain the optional input/output (I/O) cards (for interfacing digital peripherals). The front panel of the power supply is shown in Figure 3-10 (right).

The HP Model 5421A digital processor is the digital data collection and processing component of the 5400A analyzer. Its front panel is

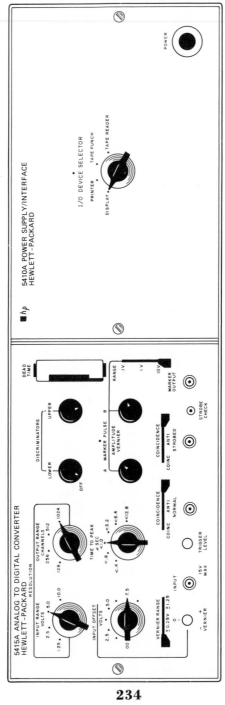

Fig. 3-10   Front panels of the HP Model 5415A ADC and 5410A power supply.

234

shown in Figure 3-11. Its controls are grouped into three main areas: data control, timing, and memory control. Under data control one of the following functions is selected: erase, accumulate, read, set-up, or transfer within memory. The accumulation modes are pulse-height analysis (PHA), sampled voltage analysis (SVA) and multichannel scaling (MCS). Data may be accumulated in either an ADD or SUB-TRACT mode. Data may be accumulated manually, automatically (erase, accumulate, read, erase, accumulate, etc.), or by single auto mode (erase, accumulate, read, stop). Under timing, dwell timing controls the counting time per channel in the MCS mode and the sampling rate in the SVA mode. A preset control allows preset time for the PHA and SVA modes and preset sweeps for the MCS mode. Under memory control, group selector *A* selects any quarter, either half, or the full memory for erase, accumulate, or transfer. Group selector *B* determines the portion of the memory (quarter, half, whole, or selected subgroup) that data are read into or out of. For transfer of data, group selector *A* determines the location in memory that a set of data is transferred from, and group selector *B* determines the location in memory that the set is transferred to. The standard memory is 1024 words, 24 bits per word. A 512-word 24-bit memory is optionally available. A unique feature of the 5400A analyzer is the "annunciator" lights used on the digital processor to key the controls pertinent to a particular mode of operation.

The HP Model 5431A Display plug-in is a unique feature of the 5400A analyzer. Designed as a plug-in for a 50-MHz main-frame oscilloscope, it gives complete flexibility over manipulation of the data displayed visually. It provides the controls for moving quarters and selecting the amount of memory displayed full scale. With the HP Model H51-180AR oscilloscope as the main frame for the 5431A display plug-in (Figure 3-12), all that is needed to have a 50-MHz oscilloscope is the HP Model 1801A dual channel vertical amplifier plug-in and the HP Model 1820A time base plug-in. Table 3-1 gives the specifications for the pulse-height-analysis mode of the HP Model 5400A Analyzer.

3-3 ANALOG-TO-DIGITAL CONVERTER (ADC) CIRCUITS

Basic ADC Circuit

The object of the ADC circuits is to provide a digital representation of the amplitude of the input signal pulses in such a form as to allow the analyzer storage circuits to tally a count in a memory channel corresponding to that digital representation. The storage control circuits can accept a number in the form of a series of short pulses, in which a number N is represented by a sequence of N pulses, separated typically by ½

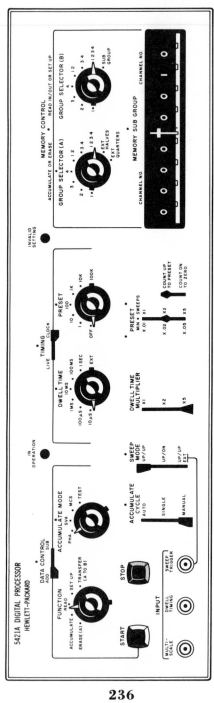

Fig. 3-11   Front panel of the HP Model 5421A digital processor.

236

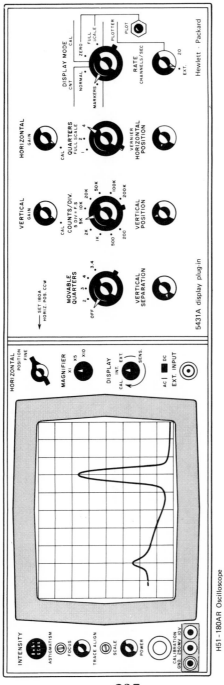

237

Fig. 3-12 Front panels of the HP Model H51-180AR oscilloscope and 5431A display plug-in.

μsec. These pulses are counted by the storage circuit "address scaler," and at the end of the pulse train the state of the scaler corresponds to the number of pulses in the train. The storage circuits tally a count in the memory channel corresponding to the state of the scaler.

Figure 3-13 shows a discharge ADC circuit in which a linear gate is included. A linear gate is an essential element of all pulse-height-to-time converters (ADC's). Its function is to admit an input signal to the converter and then to exclude all other input signals that arrive while the first signal is being measured. This prevents the latter signals from introducing errors into the measurement in progress. It follows, of course, that not every input signal is measured. However, the linear gate and the

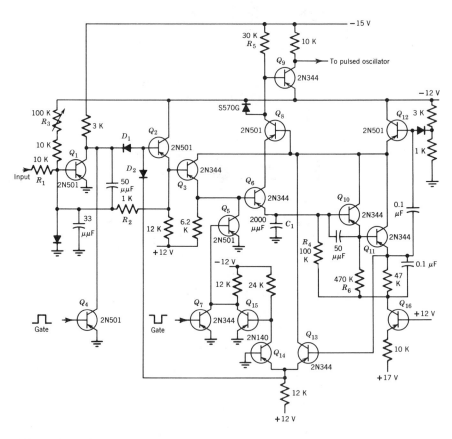

Fig. 3-13   A discharge-type pulse-height-to-time converter including a linear gate. (From R. L. Chase, *Nuclear Pulse Spectrometry,* p. 96, McGraw-Hill, New York, 1961.)

circuits that control it are designed to accept signals without prejudice with respect to their amplitudes, so that those signals which are recorded constitute a representative sample of the signals presented to the input. The spectral distribution of pulse amplitudes is, therefore, undistorted, even though some of the input signals are not measured.

In most pulse-height-to-time converters (ADC's) the linear gate is a separate and distinct circuit, but in the circuit of Figure 3-13 the linear gate shares some components with the discharge circuit, so the two must be considered together. Transistors $Q1$ and $Q2$ constitute a negative-feedback loop whose voltage gain, determined by the ratio R_2/R_1, is 0.1. Positive input signals, in the range of 0 to 100 V, appear at the emitter of $Q2$ as negative signals in the range of 0 to 10 V; R_3 is adjusted so that with no input signal present the emitter of $Q3$ is at ground potential. Normally, the circuit is in a blocked condition, with $Q4$ and $Q5$ conducting and in saturation so that even though an input signal cuts off $Q1$ no signal appears at the base of $Q6$. Shortly before the arrival of a signal to be measured, $Q4$ and $Q5$ are cut off by a positive signal at the base of $Q4$ and by a negative signal at the base of $Q7$. The base of $Q6$ remains at ground (if R_3 has been properly adjusted) until the input signal arrives and drives it negative. $Q6$ is an emitter follower responsible for charging the storage capacitor C_1 to the peak value of the signal applied to its base. Normally, a current of 0.12 mA in discharge resistor R_4 flows through $Q6$ and $Q8$, cutting off $Q9$. $Q9$ then presents a negative voltage to the pulsed oscillator (not shown), preventing oscillation. A negative signal at the base of $Q6$ causes it to charge C_1 to the signal peak. As the signal starts to decay, the base of $Q6$ is driven positive, but C_1 holds its emitter at the peak signal voltage. Thus cuts $Q6$ off so that R_5 drives $Q9$ to saturation, allowing the pulsed oscillator to start. At the same time, the current in R_4 starts to discharge C_1, Cascaded emitter followers $Q10$ and $Q11$ bootstrap the bottom end of R_4 so that the voltage across R_4, and therefore the discharge current, remain constant, leading to a linear discharge of C_1. The base current in $Q10$ constitutes a small part of the discharge current which is maintained constant by a bootstrap mechanism (see Sec. 2-2). $Q11$ bootstraps the bottom of emitter resistor R_6, in addition to discharge resistor R_4, and $Q12$ bootstraps the collector of $Q10$, so that all during the discharge $Q10$ operates at very nearly constant emitter current and collector voltage. Its base current, therefore, remains quite constant. It merely increases the discharge rate slightly without impairing its linearity. $Q12$ also bootstraps the collectors of $Q3$, $Q11$, and $Q13$, permitting the use of low-voltage transistors to handle signals larger than their collector voltage ratings.

Table 3-1

Specifications for the PHA Mode of the HP Model 5400A Analyzer

ACCUMULATION MODES

PULSE HEIGHT ANALYSIS (PHA): In this mode, the analyzer accumulates a pulse height distribution. Amplitude sorting is performed by the ADC (Analog-to-Digital Converter). Storage is in the core memory. Automatic termination of the data accumulation may be employed by presetting the accumulation time. Coarse pulse amplitude discrimination is provided. Coincidence with an externally applied signal may also be a criterion for acceptance of a pulse.

Input Pulse Requirements:

AMPLITUDE RANGE: 1.25 V; 2.5 V; 5 V; 10 V.

POLARITY: Positive.

PULSE SHAPE: > 100 ns to peak above the baseline.

INPUT IMPEDANCE: 1 kΩ, < 60 pF shunt; dc coupled.

TIME TO PEAK: 0.4 μs to 12.8 μs in binary steps. (Sets time from trigger to start of rundown.)

TRIGGER:

Sensitivity: 50 mV to 1 V adjustable (sets timing).

Distortion: Perturbs differential linearity over less than 50 mV above trigger level.

ADC Clock Rate: 100 MHz

Output Range: 128; 256; 512; 1024 channels.

Conversion Gain (Channels Out/Volt In):

RANGE: 1024 channels/1.25 volts to 128 channels/10 volts.

GAIN CHANGE ACCURACY: 2:1, ± 0.1%/step.

TEMPERATURE STABILITY: < ± 0.005%/°C.

TIME DRIFT: < ± 0.01%/24 hours.

Baseline (Input Offset):

VOLTAGE: Adjustable 0 to + 10 V in 7 steps of 1.25 V/step + vernier.

VERNIER: 0 to ± 1.25 V; 0 to ± 25 mV; OFF.

STEP ACCURACY: ± 10 mV.

COUNT RATE SHIFT: < 1 channel to 90% dead time.

TEMPERATURE STABILITY: < ± 0.1 mV/°C.

TIME DRIFT: < ± 1 mV/24 hours at fixed temperature.

COINCIDENCE STROBE: A 200 ns strobe pulse, generated at a time presettable in 400 ns intervals over the range 0-12.8 μs after the input peak time, or after the sample pulse in the SVA mode. Output monitor jack provided.

TIMING JITTER—STROBE: ± 50 ns from average.

Discriminators (UPPER and LOWER LEVEL)

RANGE: 0 to + 10 V.

Dead Time Meter:

ACCURACY: ± 5%.

LIVE TIMER ACCURACY: ± 0.5%.

Data Control: ADD or SUBtract, switched.

Timing: Count up to PRESET, or down to ZERO.

Preset Time Range: LIVE or CLOCK time, switch selectable; 0.01 minute to 5000 minutes (decade steps x multiplier in 1, 2, 5 steps).

Memory Grouping: Any quarter, any half or whole memory. Pulses exceeding selected memory range are rejected. No pulses are stored in 1st channel of group selected.

External Routing: External control of memory grouping. Routing inputs to rear panel of processor perform a function similar to the Memory Group Selector (A) control. Pulse requirements: Positive 4-12 V for 100 ns during ADC time-to-peak. Maximum pulse width less than ADC analysis time; 13 μs for 1024 channels.

Channel Capacity: To 10⁶ counts.

Memory Size:

1024 channels (standard).

512 channels (optional).

MULTICHANNEL SCALING (MCS): In this mode, the analyzer sequentially addresses each channel of the selected portion of memory and the contents of each address may be incremented by an input pulse string. Thus, each channel is used as a scaler. The SAMPLE TIME for each channel is presettable. There is no provision for coincidence or pulse amplitude discrimination. While in the Multiscale Mode there is provision for vertical display. The address information is converted to an analog voltage and available for such applications as driving a Mössbauer apparatus.

Input Pulse Requirements: (AEC Standard Compatible)
AMPLITUDE: 4-12 V.
POLARITY: Positive.
INPUT IMPEDANCE: 1 kΩ, 50 pF shunt (dc coupled).
MINIMUM PULSE WIDTH: 25 ns.
MINIMUM PULSE SEPARATION: 65 ns.

Pulse Pair Resolution: 100 ns.

Data Control: ADD or SUBtract, switched.

Timing: Count up to PRESET, or down to ZERO.

Sample Time per Channel: 10 μs to 5 s (decade steps x multiplier in 1, 2, 5 steps), or EXTernal.
EXTernal timing pulse requirements: $+4$ to $+12$ V, > 200 ns wide, > 1 kΩ, < 60 pF, Pulse Interval > 3 μs.

Preset Sweeping: 1 sweep to 500,000 sweeps (decade steps x multiplier in 1, 2, 5 steps).

Memory Grouping: Store in any quarter, half or entire memory.

Channel Capacity: To 10^6 counts.

Memory Size:
1024 channels (standard).
512 channels (optional).

Sweep Modes:
SINGLE SWEEP: PRESET SWEEP switch set to 1 (internal or external triggering).
CONTINUOUS SWEEPING: Internal or external triggering with sawtooth sweep drive; increasing channel number.
CONTINUOUS SWEEPING: Internal triggering with triangle waveform drive; increasing then decreasing channel number. (Output is available from Display Unit rear panel to drive Mössbauer apparatus for three sweep modes.)

Dead Time Between Channels: 2.2 μs.

TEST: This mode is identical with MULTICHANNEL SCALING except that input pulses are internally generated by a 1 MHz oscillator.

Baseline Checker: Closes input gate until signal level remains for 2 μs within ± trigger level. Switch selectable.

Linearity:
INTEGRAL LINEARITY: $< \pm 0.05\%$ over 100% of range.
DIFFERENTIAL LINEARITY: $< \pm 1\%$ over 100% of range. (See Trigger.)

Signal Processing Time:
PULSE ANALYSIS TIME:
Up to 128 channels and up to 3.2 μs coincidence strobe time—3.4 μs.
Up to 512 channels and up to 6.4 μs coincidence strobe time—6.6 μs.
For 1024 channels or for greater than 6.4 μs coincidence strobe time—13 μs.

SYSTEM DEAD TIME: Analysis time plus TIME TO PEAK.

System Noise (Channel Profile): Greater than 90% of the pulses from a calibrated, noise-free pulser will fall within a given channel when the pulse amplitude is set anywhere within the middle 67% of the channel. This represents less than 1 mV rms noise referred to the ADC input.

Coincidence Inputs (Normal and Strobed):
AMPLITUDE: 4-12 V.
POLARITY: Positive.
PULSE SHAPE: dc level or with specified timing.
TIMING-NORMAL: Coincidence: To enable analysis, the input must be high for > 100 ns after the pulse crosses the trigger level and prior to coincidence strobe. Anticoincidence: To disable analysis, the input must be high for > 100 ns after the pulse crosses the trigger level and prior to the coincidence strobe.
TIMING-STROBED: Coincidence: To enable analysis, the input must be high for a > 300 ns interval which includes the coincidence strobe. Anticoincidence: To disable analysis, the input must be high for a > 300 ns interval which includes the coincidence strobe.

If the discharge is to be linear, it is necessary that the reverse base-emitter leakage current in $Q6$ be negligible or, at least, constant. Therefore, during the discharge the reverse base-emitter voltage must be kept small and, preferably, constant. Diode D_2 and emitter followers $Q10$, $Q11$, and $Q13$ keep the base of $Q2$ from going more than a fraction of a volt positive with respect to the emitter of $Q6$. The base voltage of $Q6$ is determined by emitter followers $Q2$ and $Q3$ and remains about 1 volt positive with respect to the emitter during the discharge. Because the reverse base-emitter voltage of $Q6$ is small and constant, the leakage current is, likewise, small and constant. Diode D1 serves to disconnect the base of $Q2$ from the collector of $Q1$ during the discharge so that it can be successfully caught by $D2$.

Shortly after the input signal passes its peak and starts to decay, a signal from $Q9$ causes the removal of the gating signals from the bases of $Q4$ and $Q7$. The collector of $Q1$ is then once more clamped to ground potential so that subsequent input signals cannot reach $Q6$ during the discharge. However, whereas the emitter of $Q6$ is substantially negative, $Q5$ must not be allowed to be turned on because this would impose a large reverse base-emitter bias on $Q6$. This is prevented by the negative signal at the emitter of $Q13$ which turns off $Q14$, allowing $Q15$ to saturate and hold $Q5$ cut off even though $Q7$ is no longer conducting. When C_1 is nearly discharged, $Q13$ transfers current to $Q14$ which cuts off $Q15$ and turns on $Q5$, thus clamping the base of $Q6$ securely to ground potential just before the end of the discharge. As the discharge continues, $Q6$ starts to conduct once more, supplying current to R_5, cutting off $Q9$, and stopping the clock oscillator.

The zero-signal potential at the base of $Q6$ (with $Q4$ and $Q5$ not conducting) is somewhat dependent upon the base-emitter voltage drops in $Q1$ and $Q3$. However, variations with temperature in the base-emitter voltages of these two transistors affect the potential at the base of $Q6$ in the opposite sense. Of course, the emitter potential of $Q6$ varies with temperature, but this has the same effect on both the start and the end of the discharge and therefore has no net effect. The voltage across the discharge resistor R_4 depends on the potential of the emitter of $Q6$. There is, however, a compensating temperature-dependent variation in the potential of the bottom of R_4, so the slope of the discharge is not temperature dependent. This results from the fact that $Q16$ is initially in saturation so that its collector and emitter potentials are essentially the same. The base of $Q16$ is fixed at the +12-volt power supply potential while its emitter potential is free to change with temperature by about the same amount as the emitter of $Q6$. Because $Q16$ is in saturation, its collector potential changes with the emitter potential, thus maintaining the voltage across R_4 constant in spite of temperature changes. Therefore the performance of the circuit is substantially independent of ambient temperature.

Typical ADC Circuitry

LINEAR CIRCUITRY. The ADC sweep circuit and ADC logic circuits of the 512 channel analyzer Model ND120 are shown in Figures 3-14 and 3-15. These circuits generate a pulse of duration proportional to the amplitude of the input signal; this pulse is used to gate a 2-MHz oscillator in order to generate the required sequence of address scaler advance pulses (ADCA). The output pulse of the sweep circuit is the signal ADCB (called Y in the diagram; some further shaping of this signal is required elsewhere on the ADC Board. The signal Y is essentially the pulse-length modulated pulse of interest).

The mechanism for generating the length-modulated pulse, ADCB, is entirely contained in the sweep circuit of Figure 3-14. The linear gate and window-level control circuits are used for preliminary handling and control of the analyzer input pulses, which arrive at the sweep circuit pretty much in the same form as they are received from the internal or external linear pulse amplifier.

Transistors 6, 7, 8, and 9 ($Q6$, $Q7$, $Q8$, and $Q9$) comprise an ordinary feedback circuit which tends to maintain the voltage at the base of $Q7$ to be the same as that of $Q6$. The voltage at the base of $Q7$ is essentially equal to the voltage across the two parallel 1000-pF capacitors. If the voltage at the base of $Q7$ is more negative than that at the base of $Q6$, $Q9$ is cut off and the proper voltage is produced at the capacitors by means of the 5-μA current through the 10-MΩ resistor connected to +50 volts. If the base voltage of $Q7$ is higher than that of $Q6$, transistor $Q9$ conducts, and through the diodes between that transistor and the capacitors, charges the capacitors appropriately. Transistor $Q10$ normally does not conduct and therefore has no effect upon the circuit. The diode to the signal called "CLOCK" is also normally nonconducting.

A property of this feedback circuit is, obviously, that any tendency of the base voltage of $Q7$ to be more positive than that of $Q6$ is corrected quickly. If the opposite voltage error exists, however, the correction is very slow, for the diodes to $Q9$ are not capable of conducting in the direction to cause the capacitor to charge more positively. The correction is solely by means of the current through the 10-MΩ resistor.

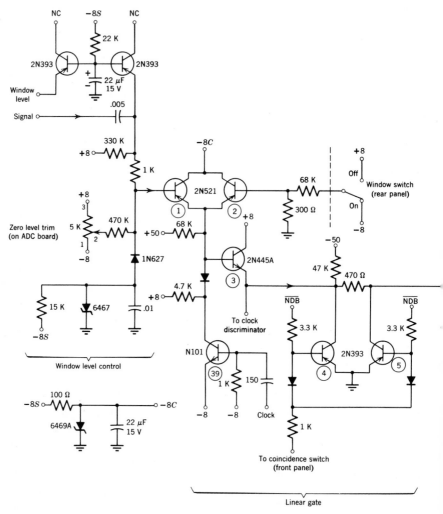

Fig. 3-14   ADC sweep circuit, linear gate and window level control (ADC Board) of the 512-channel analyzer model ND120. $\overline{\text{NDB}}$: Signal positive when analyzer is not busy.

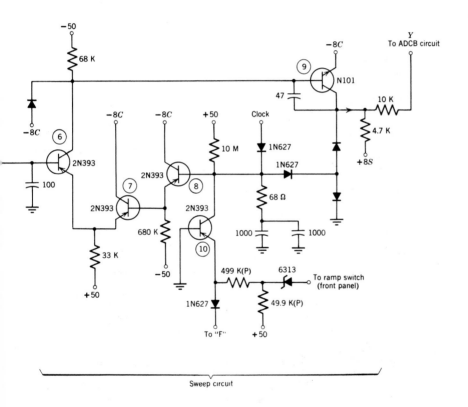

Sweep circuit

245

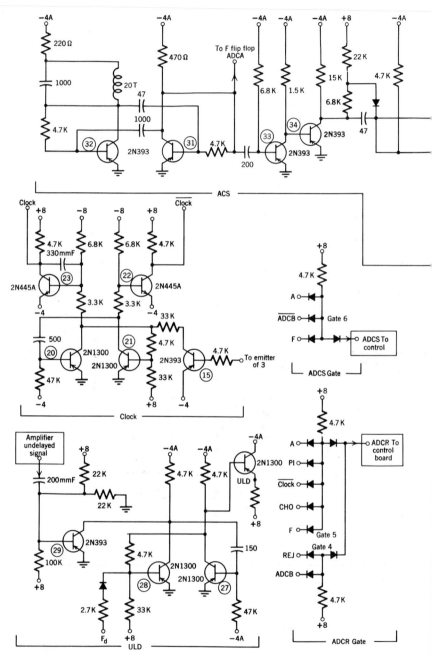

Fig. 3-15   ADC Logic Circuit (ADC Board) of the 512-channel analyzer ND120.

246

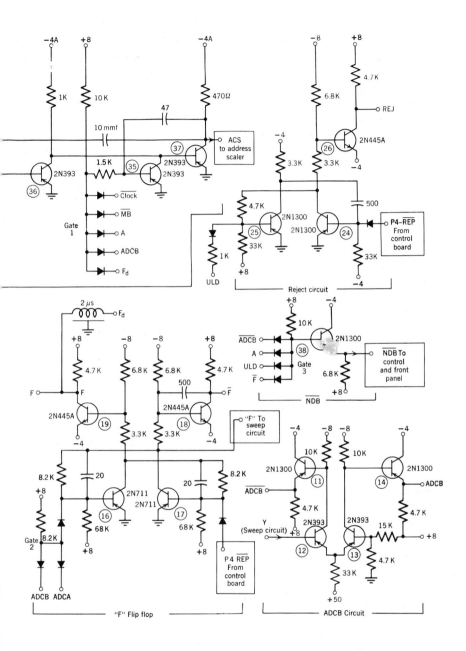

247

When an input pulse arrives, the capacitors are rapidly charged to the same voltage as the negative input pulse until the input pulse has reached a maximum value. The capacitor voltage then remains almost constant, except for the slow charging action. The voltage Y becomes several volts positive, and soon after this, transistor $Q10$ conducts (typically 60 μA), for other circuits cause the signal F to become positive, allowing current to flow through the 499K resistor. The signal called "RAMP" is a steady voltage of about 8, 16, or 32 volts depending on the setting of the analyzer "channels per 3 volts" front panel switch (Figure 3-16).

The effect of the current through $Q10$, which is very constant, and the nearly constant current through the 10-MΩ resistor, is to linearly discharge the capacitors. When they have returned to zero volts (the voltage at the base of $Q6$ is always essentially zero when the linear gate is closed, as it always is during conversion processes), the voltage Y (ADCB) quickly returns to a slightly negative value. The length of time during which Y is positive is therefore proportional to the peak voltage of the input signal pulse. The waveforms in the sweep circuit are shown in Figure 3-17.

The window-level control circuit is simply a coupling circuit that provides dc restoration for the capacitor-coupled input signals and establishes a quiescence voltage value (base line) of about zero volts (normally). The two diode-connected 2N393 transistors in this circuit comprise a temperature-compensated diode-clipping circuit which holds the baseline voltage nearly equal to the window-level voltage (from the front panel zero energy position control, see Figure 3-16) despite the action of the 330-K resistor connected to +8 volts. The trimpot shown slightly changes the voltage base line to allow long-term drifts to be compensated for, in order that the settings of the zero energy position control will be the same throughout the life of the instrument. The network diodes at the lower portion of that circuit are protective circuits that minimize the effects of possible excessively large input pulses accidentally applied to the ADC.

The linear gate circuits, transistors $Q1$, $Q2$, $Q3$, $Q4$, and $Q5$ operate as follows. Normally $Q2$ is nonconducting and therefore $Q1$ and $Q3$ are simple emitter followers. $Q39$, not actually part of the operation of this circuit, is nonconducting, so the diode between $Q39$ and $Q3$ is also nonconducting.

Any voltage pulse appearing at the base of $Q1$ also appears at the output of emitter follower $Q3$ if and only if $Q4$ and $Q5$ are nonconducting. The output of the linear gate is within a few millivolts of zero volts if $Q4$ and $Q5$ are conducting, whether an input pulse is present or not. Thus signals pass through the gate circuits either virtually "untouched," if

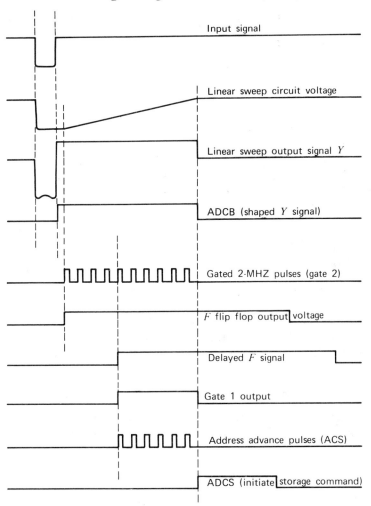

Fig. 3-17   Timing sequence of the ADC in the 512-channel analyzer ND120.

occurs while that trigger circuit is on, gate 4 produces a signal called ADCR, which causes the next memory cycle to be a nonadd memory cycle. Only if ADCB occurs very soon after a previous memory cycle will gate 4 produce a positive signal. Therefore only radiation signals occurring during the end of a memory cycle are rejected. The reason for the rejection circuits being included is to prevent events which may be distorted by the transient state of affairs at that time from being recorded, since they would be recorded incorrectly.

3-4 FERRITE CORE MEMORIES

The Residual Magnetism

In many ferromagnetic materials, the magnetic flux density B is not uniquely determined by the applied magnetic field intensity H, but depends also upon the magnetic history of the material. If H is varied cyclically, the value of B is observed to lag behind the changes in H, as indicated in the hysteresis loop of Figure 3-18. In this curve the distance "*or*" represents the residual flux density; *os*, the saturation flux density B_s; "*of*," the saturating field intensity H_s; and *oc* the coercive force. A sample of ferromagnetic material can be used as a memory element by virtue of the residual flux density. This residual magnetism can have either of two polarities, depending on the polarity of the applied magnetic field, which can be used to represent the binary digits[1] 0 and 1. Information can be stored in a piece of magnetic material if it is included in a magnetic circuit that is linked one or more times by an electrical conductor. Of the many possible geometric configurations, the most common consists of a magnetic toroid wound with one or more turns of copper wire. A current pulse in one direction in the wire writes a 1 in the core, and a current pulse in the other direction writes a zero.

To read the information stored in a magnetic core, it is necessary to determine the polarity of the residual magnetic flux. This is usually done with a read pulse that sets the core to the 0 state. The electrical signal induced in a sense winding on the core indicates whether or not this changes the residual flux. A change indicates that the core held a 1, whereas no change indicates a stored 0. The voltage induced in the sense winding is $v = n \, d\phi/dt$, where n is the number of turns in the sense

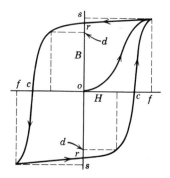

Fig. 3-18   Memory core hysteresis loop curve.

[1] If the negative residual magnetism represents 0 (or 1), then the positive represents 1 (or 0).

winding and ϕ is the magnetic flux. If the read pulse extends over a time interval Δt, then the time integral of the induced voltage is $n\,\Delta\phi$, where $\Delta\phi$ is the change in the residual flux produced by the read pulse. If the core contains a 0, the read pulse produces no net flux change, and therefore $n\,\Delta\phi = 0$. (This does not mean that no electrical signal is induced in the sense winding.) During the read pulse the flux density changes from r to s and back to r again (see the upper rsr or the lower rsr in Figure 3-18), so that $d\phi/dt$ is not equal to zero. However, the time integrals of the positive and negative parts of the induced signal are equal, so that the average value of the net voltage impulse is zero. If the core contains a 1, then switching it to 0 induces an asymmetrical signal in the sense winding which has a net integrated value proportional to the distance between the points marked r on the hysteresis loop. If the read 1 signal is to be easily distinguished from the read 0 signal, it is advantageous to use a magnetic material in which the shape of hysteresis loop is rectangular, as shown in Figure 3-18, where the distance rr is large compared with the distance rs. There is a square hysteresis loop like this in a ferrite core commonly used as the basic element of the memory.

Addressing and Sense Wires

The memory cores are generally assembled in a two-dimensional matrix of horizontal X and vertical Y wires, with a core located at each intersection (Figure 3-7). To select a particular core in the matrix, the X and Y wires threading it are pulsed simultaneously with currents, each of which is half that required to saturate the core. The selected core is thus subjected to a full saturating magnetomotive force, while other cores, which are threaded by only one of the selected addressing wires, receive only half of a saturating magnetomotive force which leaves their residual flux substantially unaltered (see Figure 3-18). The selected core can be set to either the 1 state or the 0 state, depending upon the polarity of the two addressing currents. The information stored in a core is read by pulsing that core, via its two addressing wires, to the 0 state and observing the signal in a sense wire threading the core. One sense wire going through all cores in each array (or plane) is generally connected in series to only a read (or sense) amplifier. Only one read amplifier is required per array because only one core in each array can be read at any one time.

Common Memory Arrangement

A common memory arrangement is to associate each core in a memory plane (array) with a different pulse-height channel. Each plane (array), in turn, is associated with a particular binary digit in the numbers stored in the pulse-height channels. For example, a pulse-height analyzer with 400 channels and 24 bits of storage per channel would have 24 core arrays, each with 400 cores arranged in a 20×20 matrix (Figure 3-7b). Each array is connected, via its read or sense amplifier, to a one-bit electronic register (usually a flip-flop), which collectively makes up the "memory register." The contents of a particular pulse-height channel are transferred from the core memory to the memory register by pulsing the appropriate horizontal and vertical address wires. Information is usually transferred between the core memory and the memory register in parallel; that is, all the bits of a given pulse-height channel are read simultaneously. Usually, the horizontal and vertical addressing wires of all planes (arrays) are connected in series and pulsed simultaneously. Thus, by pulsing only one horizontal wire and one vertical wire, all the bits in a selected pulse-height channel can be transferred from the core memory to a memory register. If a serial readout is used instead of a parallel readout, then each bit of the channel is read out individually.

Data for display and printout are derived from the memory register. During data acquisition the memory register participates in the process of adding 1 to the stored count in a channel whenever it is selected by the analog-to-digital converter section. This is most simply accomplished by connecting the stages of the memory register together, in sequence, with suitable coupling circuits to make it operate as a binary scaler. To tally a count in a particular channel, the contents of that channel are transferred to the memory register. A single pulse is then applied to the first stage of the register scaler, increasing the number stored by 1. The new contents of the memory register are then transferred back to the core memory. Because of its participation in the addition process, the memory register is also known as the arithmetic or "add 1" scaler.

Reading and Writing

During a read cycle, any of the selected cores switched from the 1 state into the 0 state will induce a current pulse in the sense wire for that memory plane (array). The sense amplifier with its input connected to the sense wire will set one bistable (one bit) of the arithmetic scaler. Whenever a number is read from memory, whether the count is increased by 1, as in data accumulation, or whether it is left unchanged, as in data display, it is necessary to have a write cycle to transfer the number in the arithmetic scaler back to the core memory. This results from the fact that the read operation is destructive, since in the process of reading all the cores in the selected channel are set to the 0 state. The write cycle is essentially the reversal of the read cycle. That is, the memory-core states are now to be set so that they are the same as their corresponding scaler binary states. Writing in the various planes (arrays) is individually controlled, by equipping each plane (array) with the "inhibit" wire, which threads all the cores in the plane (array) in the same sense. To write a number in a memory address, the horizontal and vertical wires corresponding to that address are each pulsed with a half saturating write 1 pulse. The current in each inhibit wire is controlled by the state of the corresponding arithmetic scaler stage, so that in those planes (arrays) in which a 1 is to be written the inhibit wire is not pulsed and the sum of two half saturating currents writes a 1. In those planes (arrays) in which a 0 is to be written, the inhibit wire is supplied with a half saturating current in the write 0 direction. The selected cores then receive two half write 1 currents and one half write 0 current. The net current, a half write 1 current, is too small to switch the selected cores, so they remain in the 0 state. The other cores in the plane (array) that share either a horizontal or a vertical address wire with the selected core receive a half write 1 current and a half write 0 current. This has no net value so those cores, of course, retain their original magnetization. The remaining cores in the plane (array) receive a half write 0 current pulse, which is too small to switch them, so they also remain in their original state.

Memory Cycle

A reset pulse sets all stages of the memory register to 0. The address scaler then directs the read current pulses to one horizontal and one vertical address wire. Each current pulse is half that required to produce core saturation and is in the direction to set the selected cores to 0. In those planes (arrays) in which the selected core contains a 1 the signals induced in the sense windings set the corresponding stages of the memory register to 1. In those planes (arrays) in which the selected core contains a 0 the signals induced in the sense windings are too small to trigger the corresponding memory register stages, so they remain in the 0 state. At this time, if the instrument is operating in the data storage mode, a pulse is applied to the input of the memory register which increases its count by 1. In the display and printout modes this step is omitted. Next, the inhibit-pulse generator directs half-saturating current pulses, in the 0 direction, to all those planes (arrays) whose memory register stage contains a 0. Shortly after the start of the inhibit pulses, and while the inhibit current continues to flow in the appropriate planes (arrays), the rewrite pulses are directed by the address scaler to the same horizontal and vertical address wires as before. The rewrite pulses are the same amplitude as the read pulses but are of opposite polarity. This causes a 1 to be written in the selected core in each plane (array) whose memory register stage contains a 1; this completes the memory cycle.

Typical Sense Amplifiers in the Memory Register

One of the typical sense amplifier is shown in Figure 3-19. The sense wire is connected to pins 42 and 44. This wire floats in the memory stack and is connected between the inputs of a differential amplifier, $Q54$ and $Q55$. Since the DC gain is less than one, and the emitters are separated with C_3, the differential amplifier has a high DC stability. The gain for the sense pulse is about 700, which results in an output pulse of about 2.5 volts, which is sufficient to drive the READ STROBE gate by means of the emitter follower $Q52$ and $Q53$. The emitters of $Q52$ and $Q53$ show a low impedance for a negative pulse and a high impedance for a positive pulse. Since the emitters are ac coupled, they will follow a negative-going pulse rather than a positive-going pulse.

$CR4$ and $R2$ form an AND gate at the base of $Q63$, with the negative sense pulse and the READ STROBE pulse as its two inputs. During the "read" time of the memory cycle, the READ STROBE is negative and will try to turn "on" $Q63$. If no sense pulse is present, the current through $R2$ is delivered by $R3$ and the voltage at the base of $Q63$ will remain unchanged. If a sense pulse is present, the anode of $CR4$ is negative and the current through $R2$ will turn $Q63$ "on." The amplitude of the current into the $Q63$ base will not depend on the amplitude of the sense pulse so long as a minimum amplitude of about 1 V is reached. The pulse at the collector of $Q63$ will set the bistable through $CR3$.

The external set signal on pin 40 through $CR1$ is connected before $CR3$ to obtain a better isolation from the bistable.

REFERENCES

Chase, R. L., *Nuclear Pulse Spectrometry*, McGraw-Hill, New York, 1961.

Strauss, M. G., ANL, Solid-State Pulse-Height Encoding System with Pileup Reduction for Counting at High Input Rates, *Rev. Sci. Instr.*, 34: 4, pp. 335–345 (April, 1963).

Instruction Manual *Operating and Maintenance Instructions for* 400 *Channel Analyzer System, Model* 20631/2, Radiation Counter Laboratories, Inc.

Chase, R. L., *Recent Developments in Multichannel Pulse-height Analysis*, BNL 5993, Brookhaven National Laboratory, Upton, N.Y., 1962.

Malmstadt, H. V., and C. G. Enke, *Electronics for Scientists*, Benjamin, New York, 1963.

Stanford, G. S., *Multichannel Pulse Analysis*, RED-EPM #120, Reactor Engineering, Argonne National Laboratory, 1963.

Strauss, M. G., ANL, Extending a Multichannel Analyzer for Two Parameter Pulse Height Analysis, *Nucl. Instr. Meth.*, 29: 1 (1964).

Millman, J., and H. Taub, *Pulse, Digital, and Switching Waveforms*, McGraw-Hill, New York, 1965.

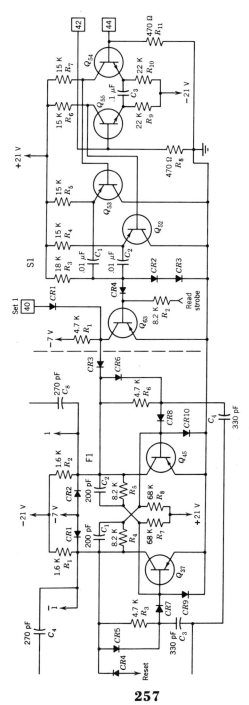

Fig. 3-19 A sense amplifier and a bistable in the RCL 400 Channel Analyzer Model 20631/2. One sense wire is connected between the input terminals "42" and "44".

257

Strauss, J. G., I. S. Sherman, R. Brenner, S. J. Rudnick, R. N. Larsen, and H. M. Mann, ANL, "High Resolution Ge(Li) Spectrometer for High Input Rates," *Rev. Sci. Instr.,* 38: 6, pp. 725–730 (June, 1967).

HP Model 5400A Multichannel Analyzer Technical Data, Nov. 1, 1967, Hewlett Packard, 1501 Page Mill Road, Palo Alto, California.

Instruction Manual for the 512 Channel Analyzer Model ND 120, Nuclear Data Inc., 3833 West Beltline, Madison, Wisconsin.

Instruction Manual for Models 402 *and* 404 "400" *Channel Pulse Height Analyzers,* Vols. I, II. Technical Instruments, Inc., 441 Washington Ave., North Haven, Conn.

Robinson, L. B., F. Gin, and F. S. Goulding, "A High-Speed 4096-Channel Analogue–Digital Converter for Pulse Height Analysis," *Nuc. Instr. Meth.* **62**, pp. 237–246 (July, 1968).

CHAPTER 4

TYPICAL NUCLEAR AND ELECTRONIC INSTRUMENTS

4-1 TYPICAL G-M TUBE SURVEY METERS

CD V-700 Survey Meter

The civil defense model V-700 survey meter in Figure 4-1 is primarily used for detecting radioactive contamination up to 50 milliroentgen/hr. The survey meter employs a beta-gamma sensitive Geiger-Müller (G-M) tube as its detector. The G-M tube,[1] housed in a nickel-plated brass probe, is connected to the survey meter (and its 900-volt dc supply) by a 3-ft cable. A rotatable shield prevents external beta particles from entering, making the detector sensitive to gamma rays only. The circuit consists of a pulse shaper, a count-rate meter, and an electronic high-voltage supply, as shown in Figure 4-2.

The pulse-shaper circuit consists of transistor $Q1$, transformer $T1$, coil $L1$, diode $D1$, and capacitor $C1$. This circuit is basically a blocking oscillator. When the instrument is "on," transistor $Q1$ is held at cutoff

---

[1] There are a collector (anode) and a cathode (wall) in the Geiger-Müller tube. The charge collected is independent of the ionization initialing it. The gas amplification (or gas multiplication, typically 10^7) increases the charge to a value that is limited by the characteristics of the tube and the external circuit.

Fig. 4-1   G-M tube survey meter CD Model V-700. (From J. G. Ello, ANL, "Radiological Survey Meters," *Electronics World*, p. 46, Jan., 1966.)

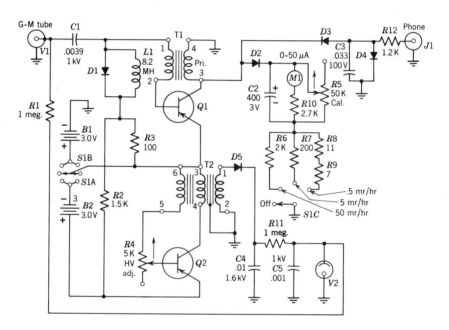

Fig. 4-2   Complete schematic diagram of the CD Model V-700 survey meter. (From J. G. Ello, ANL, "Radiological Survey Meters," *Electronics World*, p. 74, Jan., 1966.)

260

by a bias voltage network formed by resistors $R2$, $R3$, and the battery supply. The negative pulse produced by ionization in the G-M tube appears across $L1$ and $D1$. $L1$ provides a high-impedance path for these pulses, while providing a low-impedance path for direct current. The diode $D1$ prevents oscillation across $L1$. This G-M tube pulse, coupled to the base of $Q1$, activates the circuit by saturating $Q1$. At this point, most of the battery voltage $B1$ is across the winding 3-4. As the current increases in the winding, a voltage is induced in winding 1-2. This induced voltage maintains the conduction of $Q1$. In winding 3-4, the current continues to increase until the transformer core saturates. At this point, the circuit turns off, and an inductive kickback voltage or pulse appears across both windings. This pulse, appearing across winding 3-4, charges the integrating capacitor $C2$ via diode $D2$ by an amount determined by the selected range resistor $R6$, $R7$, or $R8$ and $R9$. The charge on $C2$ discharges through the meter $M1$ and its series resistor. The resulting meter current is dependent on the charge-per-pulse and pulse rate. The calibration control $R5$ shunts part of the current around the meter $M1$. The kickback voltage pulse is also used for headphone aural monitoring. This pulse is rectified by the diode $D3$ and applied to integrating capacitor $C3$. $R12$ is an isolating resistor, and $D4$ acts as a damper for the headphone.

The high-voltage supply consists of oscillator transistor $Q2$, transformer $T2$, rectifier $D5$, capacitors $C4$, $C5$, resistor $R11$, and the voltage-regulator tube $V2$, which maintains a constant 900-V output.

Before the survey meter is placed in operation, it must be calibrated against a standard source of radiation — the sole source of radiation when calibration is performed. Calibration should not be undertaken when the background reading is above normal or when in a radiation field other than that produced by the known source supplied with the instrument.

Most material is slightly radioactive. All building material, on the average, contains a few thousandths of one percent uranium. Natural radioactivity is produced by cosmic rays from outer space. The detector is constantly bombarded with this natural radioactivity. Whenever the survey meter is in the "operate" position, the meter reads this natural radioactivity. This is the activity referred to as "background." To determine the background count, set the survey meter to the × 1 position. The needle should be fluctuating over a range of 0.01 to 0.03 mR/hr. About 20 to 30 clicks per minute will be heard in the headphone. Background and other low-level radiation measurements can be made by counting the clicks and timing them with a watch.

To calibrate the CD V-700, turn the meter to the × 10 position. Present the open window of the detector probe to the center of the name plate under which is a radioactive source. The indicating meter should read between 1.5 and 2.5 mR/hr. If the indication falls below or above, adjust the calibration control R5 for an average reading of 2 mR/hr.

E-120/E-120G Geiger Counter (Survey Meter)

The Eberline E-120 (or E-120G) Geiger counter (Figure 4-3) is used for detecting radiation up to 50 mR/hr. Its circuit is shown in Figure 4-4; it consists of a high-voltage supply (Q1, Q2, Q3, T1, CR4, and V1), an amplifier (Q4), a trigger (A1), a meter driver (Q5), and a phone driver (Q6).

In the voltage supply, the oscillator transistor (Q1) drives the transformer (T1) primary and gets its positive feedback from T1's red-orange winding. The voltage is stepped up by T1's secondary, rectified, filtered, and applied to the voltage regulator V1 (type 413). V1 regulates at 900 V. The current through V1 is sensed by Q2, amplified, and used to control the current through Q3. The current through Q3 controls the bias level of Q1. This tends to hold the current through V1 to a constant value regardless of battery voltage. The result of this is that power is not wasted with new batteries, just so it will function with lower voltage batteries. This greatly extends battery life.

The amplifier Q4 amplifies and inverts the negative input pulses from the detector (G-M tube). It is biased just into cutoff, so its output is near 0 V. A pulse turns it on, and the resulting positive output pulse starts the trigger circuit (A1).

Integrated circuit A1 (MC 710G-6544A, Motorola) is connected to operate as a monostable multivibrator whose pulse width is controlled by the RC time constant between its pins 7 and 3. This time constant is established by the setting of S1B (scale selection) which selects a particular R and C. The calibration controls form the R for each scale, making the pulse width continuously adjustable for calibration. When the trigger is initiated by the pulse from Q4, the output at pin 6 goes positive and holds until the predetermined time (RC) elapses.

The meter driver Q5 is normally off, so no current flows through M1. When the trigger is on, Q5 is turned on and current flows. The amount of current is determined by the voltage on the base of Q5 and R15. The length of time that current flows is determined by the pulse width of the trigger. This (current times time) forms a certain charge which is transferred to C9 for each event counted. C9 discharges through M1, yielding a certain average current dependent on the rate of input pulses. Changing the pulse width of the trigger (i.e., changing scales or calibration pot setting) changes the average current for a given input pulse rate. This allows the meter to be calibrated to read in mR/hr (or cpm) at the detector.

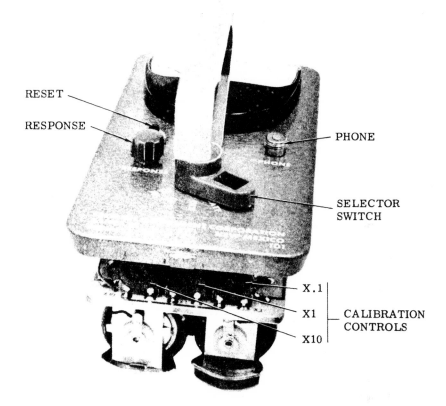

Fig. 4-3   Eberline Model E-120/E-120G geiger counter, showing the internal and external controls.

The response time of $M1$ is controlled by the RC time constant of $C9$ and $R16$, the response control. With $R16$ set to low resistance the time constant is fast, and at high resistance it is slow.

The phone driver $Q6$ amplifies and inverts the output pulse from the trigger, yielding a large amplitude negative going pulse which is capacitively coupled to the PHONE connector.

4-2   ION-CHAMBER-TYPE SURVEY METER MODEL 440

General Description

The Victoreen Model 440 portable survey meter is an ion-chamber instrument designed to measure gamma radiation over a broad energy

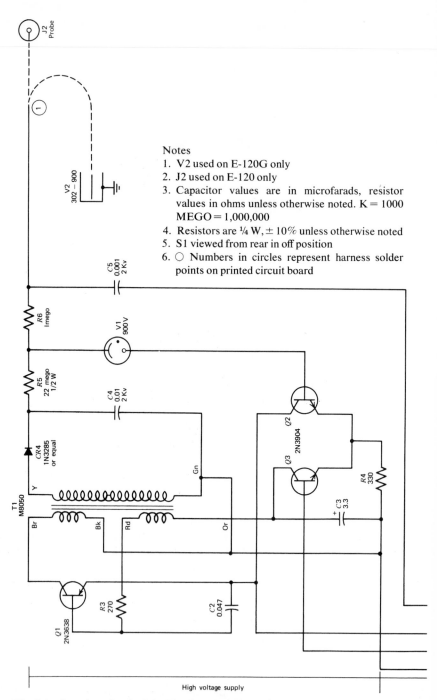

Notes
1. V2 used on E-120G only
2. J2 used on E-120 only
3. Capacitor values are in microfarads, resistor values in ohms unless otherwise noted. K = 1000 MEGO = 1,000,000
4. Resistors are ¼ W, ± 10% unless otherwise noted
5. S1 viewed from rear in off position
6. ◯ Numbers in circles represent harness solder points on printed circuit board

High voltage supply

Fig. 4-4   Complete circuit of the Eberline Model E-120/E-120G G-M Counter.

264

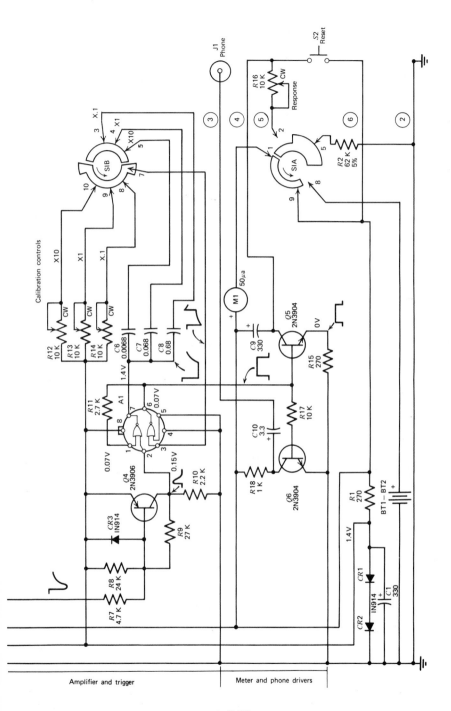

Calibration controls

Amplifier and trigger | Meter and phone drivers

265

range and at extremely low intensities. The energy range is 30 keV to 1.3 MeV $\pm$ 10 percent with end cap and 6.5 keV to 1.3 MeV $\pm$ 15 percent with end cap removed. There are five intensity ranges, namely, 0–3, 0–10, 0–30, 0–100, and 0–300 mR/hr. The accuracy is $\pm$ 10 percent of true dose rate at midspectral response.

The ion chamber[1] is an air nonsealed chamber; as a consequence a barometric and temperature correction is necessary where extreme accuracy is desired. The cylindrical walls of the chamber are constructed of $\frac{1}{16}$ in. paper-base phenolic to provide maximum strength and minimum gamma absorption. The front of the chamber is constructed of two thicknesses of $\frac{1}{4}$ mil mylar electrically isolated from each other. The inner wall of the phenolic tube and inner mylar film are coated with aquadag and the ion-collecting potential is applied to this surface. The outside of the phenolic tube and the outer mylar film are coated with a conducting material and grounded to provide adequate electrostatic shielding for the inner conducting surface. Also, the space between the two mylar films is well vented to the atmosphere so that displacements to the outer window are not transmitted to the inner where they would otherwise result in severe transients. Since the front window is thin mylar the chamber will have good beta response, the minimum beta energy being about 40 keV. The aluminum end cap is supplied to provide some beta discrimination.

The block diagram is shown in Figure 4-5. When the ion chamber is exposed to radiation, a current is caused to flow through the "Hi-Meg" resistor $R1$, developing a voltage drop between the input and output terminals of the system. This voltage is impressed across the dynamic capacitor $C111$. The capacitance of $C111$ is caused to vary by a magnetic circuit comprised of one plate of $C111$ and $L1$ and driven at a frequency of 300 Hz by an oscillator. The voltage impressed on $C111$ is thus varied at 300 Hz and is coupled to an ac amplifier through the coupling capacitor $C211$. After having been amplified by the ac amplifier, the ac voltage is applied to a discriminator circuit, where its phase is compared to the phase of the oscillator voltage. The resultant voltage is detected and used as a feedback voltage. The amount of feedback is determined by the resistance attenuator and hence determines the range of the instrument.

Circuitry

DYNAMIC CAPACITOR   Figure 4-6 shows the complete circuit of the Victoreen Model 440. The dynamic capacitor ($C2$) consists of four major

---

[1] An ionization chamber is normally operated in the region of the saturation current. The gas amplification is unity and the instrument response is nearly proportional to the ionization produced in the chamber.

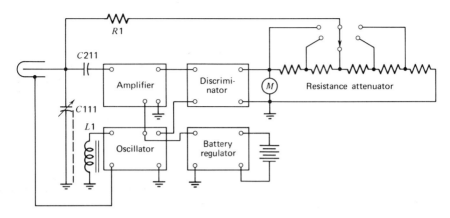

Fig. 4-5   Block diagram of Model 440.

elements: a variable capacitance to ground, a low leakage coupling capacitor, a drive coil, and a pickup coil.

The operating frequency of the reed element is controlled by the mechanical resonance of the reed. Operation is as follows. The drive coil which is between terminals C and B of the probe is driven by the transistor $Q7$ which causes the reed to move. A pickup coil, which is physically located on the opposite side of the reed, has a permanent magnet located in the center of the coil. (This is the coil between terminals A and B). When the reed moves, a voltage is induced in the pickup coil and is applied to the base of $Q6$. This voltage is amplified by $Q6$ and $Q7$ and appears at the drive coil in a polarity such as to sustain the motion. The reed will continue to move until transistor $Q7$ is cut off. At this time the direction of motion reverses, and the cycle motion causes the reed to oscillate. Since the motion of the reed is maximum at its self-resonant frequency, the system will operate at that frequency.

AC AMPLIFIER.   The current from the ion chamber impresses a charge on the dynamic capacitor plate. This charge is converted to an ac voltage and coupled through the coupling capacitor to the grid of the ac amplifier tube $V1$. The voltage is amplified by $V1$ and coupled through a step-down transformer to the base of $Q1$. The signal voltage is amplified by $Q1$, $Q2$, and $Q3$, which operate as a 3-stage cascade amplifier. The voltage at the collector of $Q3$ is transformer-coupled to the phase discriminator, which consists essentially of $T1$, $CR5$, $CR6$, and $T2$.

DISCRIMINATOR.   In order that the output circuit may be able to sense input dc polarity, a phase discriminator is incorporated in the circuitry;

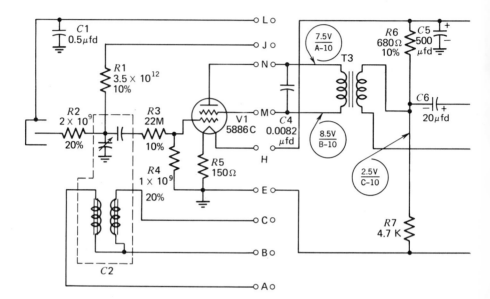

Fig. 4-6a   Amplifier section of the Victoreen Model 440 Survey meter. For the complete circuit, this section is connected to that shown in Figure 4-6b by Line 1 and Line 2.

All voltages measured with 20,000 Ω/V meter, range switch in 300 MR/HR position and no significant radiation field present
All indicated voltages (+) unless noted otherwise

it operates as follows. When a dc voltage is impressed across the dynamic capacitor, an ac voltage appears at the secondary of $T2$, the phase of the voltage depending on the polarity of the impressed voltage and the instantaneous position of dynamic capacitor.

A reference-phase voltage is supplied to the discriminator by the oscillator that drives the dynamic capacitor through transformer $T1$. Transformer $T2$ is connected to the center tap of $T1$ and supplies a voltage to diodes $CR5$ and $CR6$, whose amplitudes depend on the phase of the voltages between $T1$ and $T2$. If one side of $T1$ is adding to $T2$, then the other side will be subtracting from $T2$ and the dc output voltage on one side of the discriminator will increase and the other side will decrease. If the input voltage to the vibrating reed is reversed, the instantaneous output voltage of $T2$ will be reversed, causing the procedure to be reversed.

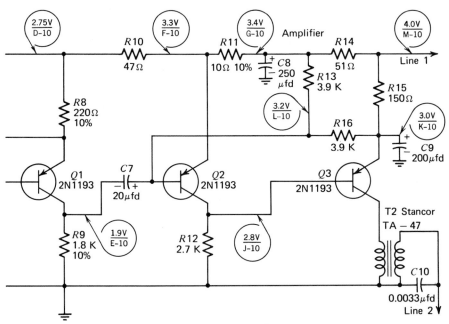

Notes for Figs. 4-6*a* and *b*:

Voltage
Position – VM range

POWER-SUPPLY REGULATOR. The power-supply regulator consists of *Q*4, *Q*5, and *CR*1, connected in the conventional series-type regulator, with *CR*1 being used as the reference voltage. A portion of the input voltage is coupled to the base of *Q*4 through *R*17 to compensate for variations in input battery change.

4-3 GAS-PROPORTIONAL ALPHA-COUNTER MODEL PAC-4G

General Description

The Eberline Model PAC-4G proportional counter (Figure 4-7) is a portable, battery-operated instrument used for the detection and measurement of alpha radiation. It has a gas-flow proportional-type detector, and the count rate is read out by the Eberline LIN-LOG presentation. This presentation eliminates all scale switching and multiplying factors, yet retains linear increments within each decade. Four decades are covered using two meter movements to yield a scale length of about 1.5 in. per

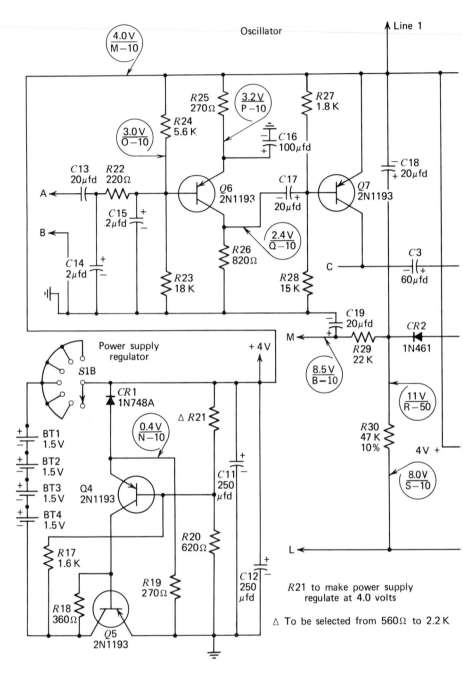

Fig. 4-6*b*   Oscillator, discriminator and power supply regulator of the Victoreen Model 440.

270

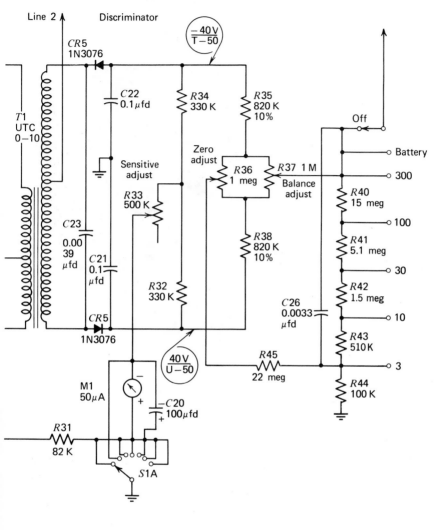

Note for Figs. 4–6a and b:
All resistors are ½W 5% tolerance
unless otherwise noted

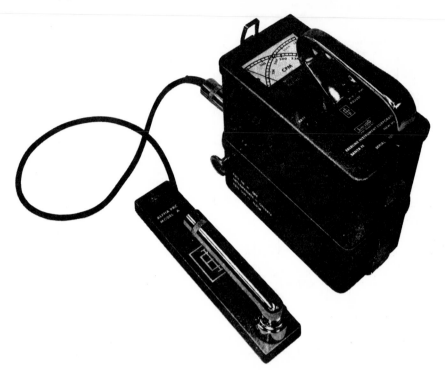

Fig. 4-7   Eberline Model PAC-4G Gas Proportional Alpha Counter.

decade, which gives maximum readability and ease of interpretation. Full-scale reading for four decades are 500, 5 K, 50 K,and 500 K counts per minute, respectively. Calibration is such that the meter reads 2π value of a 1-in. diameter Pu^{239} source.

Specifications of the detector (EIC Model AC-21 or AC-21B) are as follows:

1. Active area: 50 cm² (7.7 in.²) within 7.2 × 1.3 in. sampling area.
2. Window thickness: 0.85 mg/cm² aluminized mylar.
3. Efficiency: 50 percent of 2π for alphas emitted from 50 cm² of a distributed area Pu^{239} source.
4. Uniformity: no single reading from a 1-in. diameter Pu^{239} source deviates more than ± 5 percent from the average reading.
5. Operating voltage: in the range of 1400 to 2000 V.
6. Plateau: with Pu^{239} source typically 300 volts long, with slope less than 3 percent/100 V.
7. Gas: pure grade propane.
8. Size: approximately 9 in. long × 2 in. wide × 1½ in. high.

The detector, when in a radioactive field, generates electrical pulses. These pulses are coupled into an amplifier, and after amplification they are applied to a pulse-height discriminator. If they are of sufficient amplitude, they cause the binary to change states. The square-wave output of the binary is applied to the driver. This converts the square-wave frequency into a proportional dc current which deflects the meters. Thus, the meter reading is proportional to the intensity of radioactive field at the detector. The complete circuit of the Model PAC-4G Counter is shown in Figure 4-8.

Functional Theory

DETECTOR. When an alpha or beta particle penetrates the window and passes through the gas, it ionizes some of the gas molecules. The negative ions generated are accelerated toward the anode wire due to the positive high voltage applied. As they approach the anode, these ions will ionize more molecules, resulting in gas amplification. The amount of amplification is controlled by the high voltage applied to the detector. The group of ions are collected, and the resulting current pulse is coupled through the cable into the instrument, then into the amplifier. The output pulse is directly proportional to the primary ionization for a given voltage.

AMP-DISCR-BINARY CARD. (P-202). The amplifier portion of this card consists of $Q201$ through $Q205$, together with associated circuitry. $Q201$ and $Q202$ form a differential amplifier input. $Q203$ and $Q204$ are direct-coupled gain stages. $Q205$ is basically an emitter follower with a feedback signal taken from the emitter. The amount of feedback applied to $Q202$ base is controlled by the ratio of resistances of $R207$ and $R205$. The higher the resistance of $R207$, the more feedback there is, thereby decreasing the gain.

The discriminator consists of $CR202$ and $Q206$. The collector of $Q205$ is a current source which drives tunnel diode $CR202$. When the current, due to a pulse, exceeds the diode peak current, it switches to its high voltage state. When it switches, it saturates $Q206$, causing a sharp decrease in its collector voltage. After the pulse $CR202$ switches back to its low voltage state, $Q206$ is turned off, and its collector is free to rise back to positive voltage.

The binary consists of $Q207$ through $Q210$. The two stable states of the binary are as follows: (a) $Q207$ and $Q210$ "ON," $Q208$ and $Q209$ "OFF," resulting in output H at 0 V and output J at + supply voltage; (b) $Q208$ and $Q209$ "ON," $Q207$ and $Q210$ "OFF," resulting in output H at + supply voltage and output J at 0 V. These two states are entirely symmetrical and the binary holds either state due to cross connections from outputs to opposite bases. Switching from one state to the other

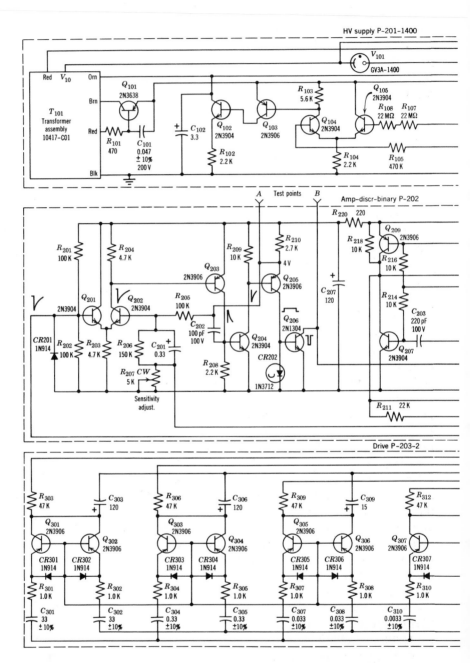

Fig. 4-8   Complete circuit of Eberline Model PAC-4G Proportional Alpha Counter.

274

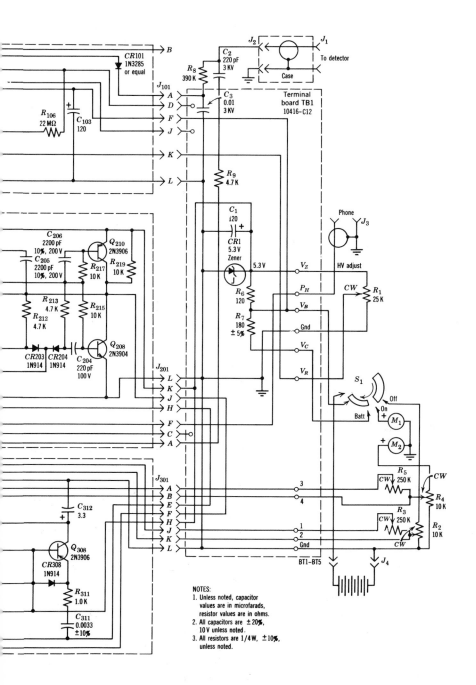

NOTES:
1. Unless noted, capacitor values are in microfarads, resistor values are in ohms.
2. All capacitors are ±20%, 10 V unless noted.
3. All resistors are 1/4 W, ±10%, unless noted.

275

occurs when $Q206$ saturates. This signal is coupled through the forward-biased diode ($CR203$ or $CR204$), and through the corresponding capacitor ($C203$ or $C204$) into the base of the ON transistor ($Q207$ or $Q208$). This turns that transistor OFF, and its corresponding rise in collector voltage is coupled to the other side to switch the binary. Thus the two outputs (pins H and J) are alternately at + supply voltage and ground switching states for each pulse that exceeds the discrimination level.

DRIVER CARD (P-203).   The driver card contains four similar sections, one for each decade of frequency response, which convert the signal from the binary into a dc current which is proportional to the frequency of the input. The pair of input capacitors ($C301$ and $C302$ for section 1, etc.) determine the frequency response for each section. These capacitors are different by a factor of 10 in each section, causing the same difference in the frequency response. The capacitors on the output collector of each section ($C303$, etc.) slow the meter response time to aid in averaging the meter reading.

CALIBRATION.   Sections 1 and 2 of the driver combine to deflect meter one and sections 3 and 4 combine to deflect meter two. The calibration controls adjust this combination for the proper meter reading versus input count rate. Each pair of sections and meter is completely independent of the other. The first control ($R2$ and $R4$, respectively) adjusts the slope of the meter current versus frequency line for both sections. The second control ($R3$ and $R5$, respectively) adjusts the point at which the lower frequency section saturates. This saturation forms a stable base upon which the higher-frequency section operates.

HIGH-VOLTAGE POWER SUPPLY (P-201).   This unit consists of a controlled oscillator ($Q101$), voltage step-up and rectification (10417-$C01$), high-voltage reference ($V101$), differential amplifier sensing and reference input ($Q104$, $Q105$), amplifier ($Q103$), and oscillator control ($Q102$). The supply regulates on voltage above the regulator tube ($V101$) and is made variable if the reference input is changed. As the reference is increased, the differential amplifier is unbalanced, which increases the level of the oscillator and the high voltage output. This puts more current into the sensing input of the differential amplifier, rebalancing it at a higher output voltage.

Changes in high-voltage load or battery voltage are regulated as follows. A heavier load or lower battery will tend to decrease the high voltage, unbalancing the differential amplifier. This increases the level of the oscillator in order to maintain the original high-voltage output. More current is thereby drawn from the battery.

Since the supply regulates on voltage above the regulator tube, the range of output voltages can be changed simply by selecting a tube with a different voltage rating.

4-4 EXPONENTIAL PULSE GENERATOR MODEL PG-75

The exponential pulse is often used as an input to instruments for testing linearity and waveform shaping. Figure 4-9 is the circuit of the exponential pulse generator Model PG-75. The output voltage from the power supply is either positive or negative, which is controlled by the POLARITY switch. The ATTENUATION switch provides the two factors 1 and 10, with the 10-step control. For the factor 1 (or 10), 1 (or 0.1) V per step is given. The vibrating mercury relay (CP Clare, HGP-2028) operates at 60 Hz and is used to control the charge and discharge rate of the decay time capacitors, so that the exponential output of the *RC* network is at the rate of 60 Hz. The decay time is determined by the decay-time capacitor, 1 K resistor, and load.

4-5 TYPE 321 OSCILLOSCOPE

General Description

INTRODUCTION. The Tektronix Type 321 in Figure 4-10 is a transistorized, battery-operated portable oscilloscope. Its vertical deflection system has a bandpass from dc to 5 MHz, a rise time of 0.07 μsec, a sensitivity of 0.01 to 20 V/div in 11 calibrated steps, and an input impedance of 40 pF, paralleled by 1 MΩ, etc. The sweep rates are from 0.5 μsec/div to 0.5 sec/div in 19 calibrated steps. With the external horizontal input, the bandpass is dc to 1 MHz, the sensitivity is 1.5 V/div with sweep magnifier, and the input impedance is 20 pF, paralleled by 100 kΩ.

The basic principle for any type of oscilloscope is the same, and so its circuitry may be divided into the six sections: vertical amplifier, triggering, sweep generation, horizontal amplifier, cathode-ray tube (CRT) circuit, and power supply.

BLOCK DIAGRAM. Figure 4-11 is the block diagram of the type 321. Waveforms to be displayed are connected to the vertical INPUT terminal. Large signals are attenuated the desired amount (up to 2000 times) in the attenuator networks. The signal is then amplified in the vertical amplifier and fed push-pull to the vertical deflection plates of the CRT.

The trigger-pickoff circuit in the vertical amplifier applies a sample of the input signal to the time-base trigger circuit. This sample signal is

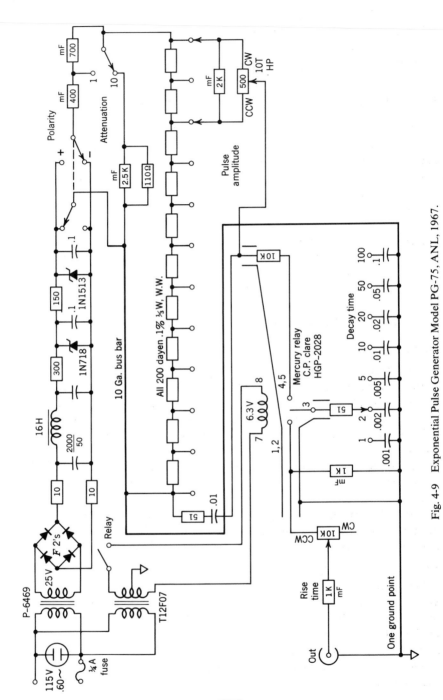

Fig. 4-9   Exponential Pulse Generator Model PG-75, ANL, 1967.

278

FOCUS—Controls sharpness of spot or trace.

ASTIGMATISM—Used in conjunction with FOCUS to obtain overall focus.

VERTICAL CONTROLS

VERTICAL POSITION—Positions trace vertically.

AC-DC—Selects either AC or DC input coupling.

VOLTS/DIV. and VARIABLE—Selects vertical deflection factor.

INPUT—Terminal for accepting waveforms to be displayed on CRT.

CAL OUT 500 MV.—Terminal provides 500-mv square wave for compensating probe.

DC BAL.—Potentiometer for setting dc balance of Vertical Amplifier.

INPUT—Terminal for accepting external triggering signal.

INTENSITY—Controls brightness of trace.

SCALE ILLUM.—Adjusts brightness of graticule markings (when operating from AC line).

HORIZONTAL CONTROLS

HORIZONTAL POSITION—Positions trace horizontally.

POWER—Switch turns power on and off.

TIME/DIV. and VARIABLE—Selects sweep rate.

EXT. HORIZ. INPUT—Terminal for accepting external horizontal signal.

TRIGGERING CONTROLS

LEVEL—Selects point on triggering signal at which sweep is triggered.

SLOPE—Determines whether sweep is triggered on + or − slope of triggering signal.

AC-DC—Selects AC or DC coupling for triggering signal.

INT-EXT—Selects either internal or external triggering signal.

STABILITY—Potentiometer for setting dc level of sweep generator.

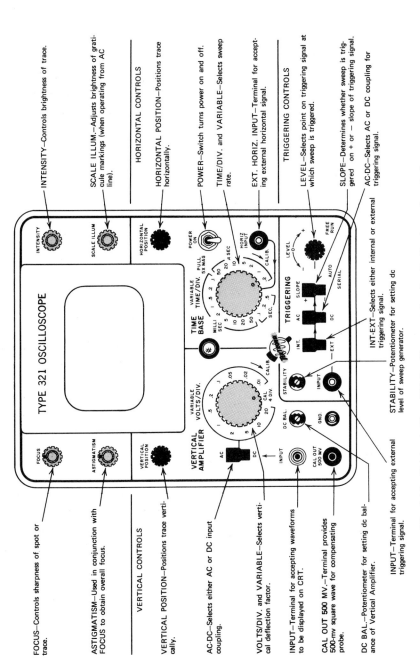

Fig. 4-10 Tektronix Type 321 Oscilloscope, showing the front-panel controls.

279

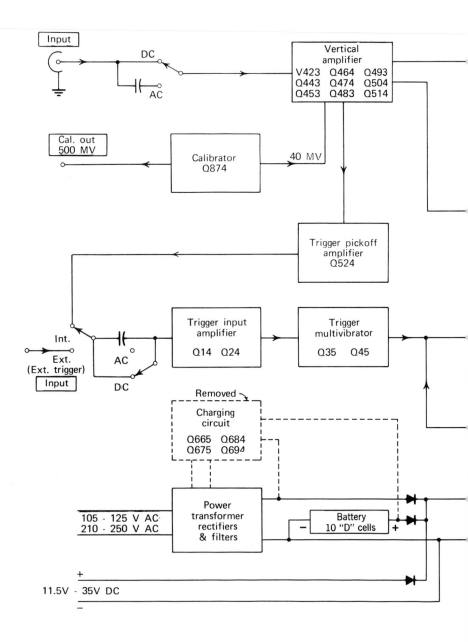

Fig. 4-11   Block diagram of the Tektronix Type 321 Oscilloscope.

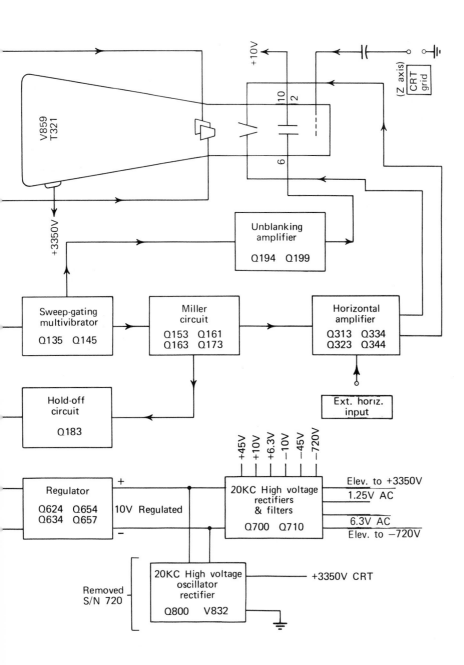

V859
T321

+3350V

+10V

10
2

6

(Z axis)
CRT
grid

Unblanking
amplifier

Q194   Q199

Sweep-gating
multivibrator

Q135   Q145

Miller
circuit

Q153   Q161
Q163   Q173

Horizontal
amplifier

Q313   Q334
Q323   Q344

Hold-off
circuit

Q183

Ext. horiz.
input

+45V
+10V
+6.3V
−10V
−45V
−720V

Regulator

Q624   Q654
Q634   Q657

+

10V Regulated

−

20KC High voltage
rectifiers
& filters

Q700   Q710

Elev. to +3350V

1.25V AC

6.3V AC

Elev. to −720V

Removed
S/N 720

20KC High voltage
oscillator
rectifier

Q800   V832

+3350V CRT

281

instrumental in starting the horizontal sweep. An external triggering signal, connected to the trigger INPUT terminal, may also be used for this purpose. Signals of widely varying shapes and amplitudes may be applied to the time-base trigger circuit. This circuit in turn produces constant-amplitude output pulses which are used to start the horizontal sweep at the proper time to ensure a stable display of the vertical-input waveform.

The output pulses from the time-base trigger circuit are applied to the time-base generator to initiate the sawtooth horizontal-sweep waveform. The sawtooth waveform is then amplified in the horizontal amplifier and applied push-pull to the horizontal deflection plates of the CRT. For *XY* applications of the instrument an externally generated signal can be applied to the EXT. HORIZ. INPUT terminal. The external signal is then amplified by the horizontal amplifier and applied to the horizontal deflection plates of the CRT.

The calibrator produces a square-wave output of constant amplitude which can be used to check the gain of the vertical amplifier and compensate the probe. A 40-mV square wave, peak-to-peak, is coupled internally to the vertical amplifier when the VOLTS/DIV. switch is set to CAL. 4 DIV. position.

Vertical Amplifier

The vertical amplifier circuit is shown in Figure 4-12.

INPUT CIRCUIT. The vertical amplifier requires an input signal of 0.01 V, peak-to-peak, to produce one division of calibrated deflection on the cathode-ray tube. To satisfy this condition, while making the instrument applicable to a wide range of input voltages, a precision attenuation network is incorporated into the vertical deflection system.

The attenuators are shown in Figure 4-13. They are frequency-compensated voltage dividers. For dc and low-frequency signals they are resistance dividers, and the degree of attenuation is determined by the resistance values. The impedance of the capacitors, at ac and low frequencies, is so high that their effect in the circuit is negligible. As the frequency of the input signals increases, however, the impedance of the capacitors decreases and their effect in the circuit becomes more pronounced, For high-frequency signals the impedance of the capacitors is so low, compared with the resistance in the circuit, that the attenuators become capacitance dividers.

INPUT CATHODE FOLLOWER AND EMITTER FOLLOWER. A nuvistor, V423, is used as a cathode-follower input stage. This stage presents a high-impedance, low-capacitance load to the input circuit and isolates the input circuit from the main amplifier. An emitter-follower stage $Q443$

couples the input cathode-follower to the input amplifier. The output impedance of $Q443$ is very nearly equal to the output impedance of $V423$ divided by the β of $Q443$. This EF stage thus provides the necessary low-impedance drive (approximately 20 Ω) for the base of $Q464$, one-half of the input amplifier stage. The opposite EF $Q453$ couples a dc voltage, adjustable by means of the DC BAL. control, to the base of $Q474$ (the other half of the input amplifier stage).

INPUT AMPLIFIER AND SECOND EMITTER FOLLOWER.   The input amplifier consists of $Q464$ and $Q474$ connected as an emitter-coupled paraphase amplifier. In addition to amplifying the signal, this stage converts the single-ended input at the base of $Q464$ to a push-pull output signal between the two collector circuits. $Q474$ operates essentially as a grounded-base amplifier (grounded through the low output impedance of $Q453$); the input signal to $Q474$ is developed across the impedance in its emitter circuit. The VARIABLE control $R478$ and the GAIN ADJ. $R468$ vary the emitter degeneration and thus affect the gain of the input amplifier. The DC BAL control $R432$ is used to adjust the dc level of $Q474$ so that its emitter will be at the same voltage as the emitter of $Q464$ when no input signal is applied to the instrument. The VERTICAL POSITION control is a dual control, connected between $+10$ and -10 V. It is connected electrically, so that as the voltage at one arm changes in the positive direction the voltage at the other arm changes in the negative direction.

The second EF stage ($Q483$ and $Q493$) provides a high-impedance, low-capacitance load for the input amplifier and provides a low-impedance drive for the base of the output amplifier.

OUTPUT AMPLIFIER.   The output amplifier $Q504$-$Q514$ is a conventional collector-loaded, push-pull amplifier that is used to drive the vertical deflection plates of the CRT. There are two time-constant networks connected between the two emitters. One is the network consisting of $R507$, $R508$, $C507$, and $C508$. This is an extremely short time-constant network (a fraction of a microsecond) and affects only fast-rise signals. The capacitive branch of this network offers less degeneration at high frequencies and thus improves the high-frequency response of the stage. The amount of high-frequency compensation can be adjusted by means of variable capacitor $C508$.

The other time-constant network consists of $R517$, $R518$, and $C518$. This is a much longer time-constant network, and compensates for the thermal time-constant of the transistors. At direct current and extremely low frequencies the impedance of $C518$ is so high that the effect of this network is negligible. Above this range, however, the impedance of $C518$

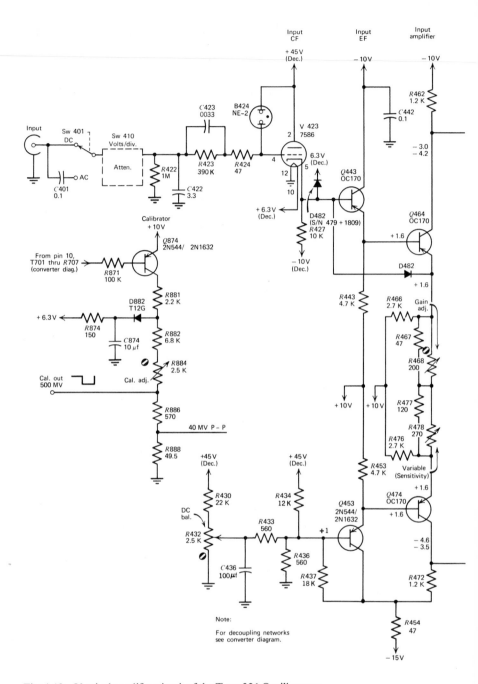

Fig. 4-12 Vertical amplifier circuit of the Type 321 Oscilloscope.

Voltage readings were obtained with controls set as follows:
Input signal . . . . . . none

284

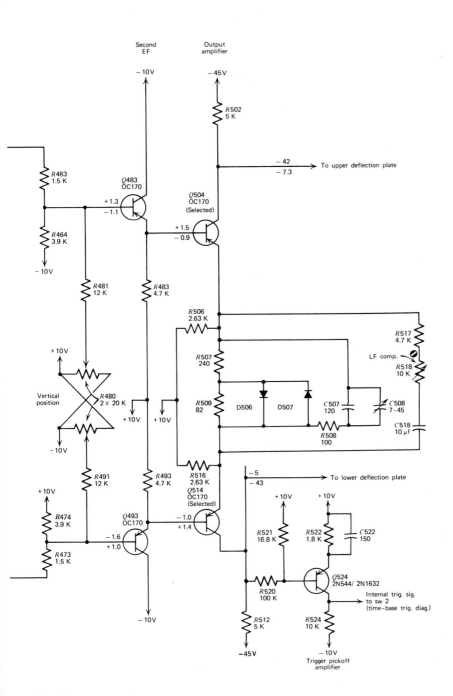

Volts/div. 1
Vertical position . . .
Full left (upper reading)
Full right (lower reading)

285

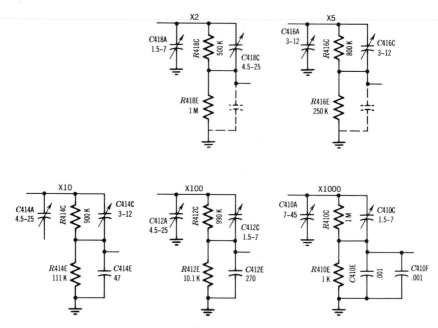

Fig. 4-13   Attenuators connected to the input circuit of the vertical amplifier.

decreases and the shunting resistance of $R517$ and $R518$ decreases the degeneration slightly. $R518$, the LF COMP. control, is used to adjust the time constant for optimum results.

The dc level of the output amplifier is established, in part, by the divider between the collector of the input amplifier and the base of the second EF ($R463$ and $R464$ on one side and $R473$ and $R474$ on the other). These dividers help set the dc level of $Q504$ and $Q514$ so that the maximum swing in collector voltage can be obtained.

Triggering

TIME-BASE TRIGGER.   The time-base trigger circuit in Figure 4-14 consists of a trigger input amplifier stage $Q14$-$Q24$ and a rectangular-pulse trigger multivibrator $Q35$-$Q45$. The function of the trigger circuitry is to produce a positive-going rectangular pulse at the collector of $Q45$ whose repetition rate is the same as that of the triggering signal. The positive step is then differentiated to produce a very sharp positive spike (trigger) which is used to trigger or start the sweep.

The triggering input signal from which the rectangular output is produced may be obtained from an external source through the (TRIGGER-

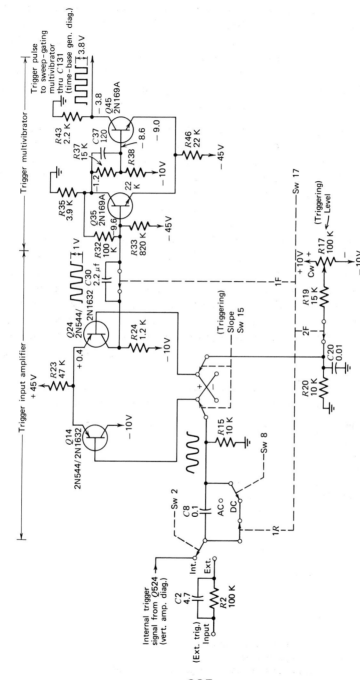

Fig. 4-14   Time-base trigger circuit of the Type 321 Oscilloscope.

287

ING) INPUT terminal, or it may be obtained from the vertical amplifier. When the INT-EXT. switch SW2 is in the INT. position, the signal is received internally from the vertical amplifier. A trigger-pickoff stage $Q524$ (shown in Figure 4-12) receives a sample of the vertical-output signal from the collector circuit of $Q514$. The amplified signal at the collector of $Q524$, which is in phase with the vertical input signal at the grid of $V423$, is then coupled to the time-base trigger circuit.

TRIGGER MULTIVIBRATOR.   The trigger multivibrator is a bistable Schmitt circuit. It is forced from one of its stable states into the other by the triggering signal applied to the base of $Q35$. In the first stable state (ready to receive a signal) $Q35$ is cut off and its collector voltage is up (near ground). This holds up the base of $Q45$, since the two circuits are dc-coupled, and $Q45$ conducts. With $Q45$ conducting, its collector voltage is down; hence no output is being produced. A positive-going signal is required at the base of $Q35$ to force the multivibrator into its second stable state to produce the positive step at the collector of $Q45$. However, since the signal at the base of $Q35$ is an amplification of the triggering signal, it contains both positive- and negative-going portions. The positive-going portion of the triggering signal will drive the base of $Q35$ in the positive direction. When the base voltage reaches cutoff, $Q35$ starts conducting and its emitter voltage will rise, following the base. This pulls up the emitter of $Q45$, since the two emitters are strapped

---

Waveforms and voltage readings were obtained with controls set as follows:
Waveforms
Input signal. . . . . 500 $\simeq$ sinewave, 3 V P.-P.
Trigger source . . . . int.
Trigger slope. . . . . +
Trigger level . . . . auto
Time/div. . . . . 0.5 msec
Variable (time/div.) . . . . calib.
Mag.. . . . . off
Voltage readings
Trigger slope. . . . . +
Input signal. . . . . none
Trigger level . . . .
clockwise, but not switched
to free run

Note for Sw17
① Ganged with R17, triggering level control
② Open only in auto. triggering position
③ For additional switch details see time-base generator diagram

together. At the same time the collector voltage of $Q35$ starts to drop, which pulls down the base voltage of $Q45$. With the base of $Q45$ down and its emitter up, $Q45$ cuts off. As $Q45$ cuts off, its collector voltage rises, creating a positive step at the output. This transition occurs very rapidly, regardless of how slowly the base signal of $Q35$ may rise. When the signal at the base of $Q35$ starts in the negative direction, just the opposite chain of events will occur; $Q35$ will cut off and its collector voltage will rise. This will pull the base $Q45$ out of cutoff and $Q45$ will conduct. As $Q45$ conducts, its collector voltage drops; this completes the positive step-voltage output from the trigger multivibrator circuit.

TRIGGER-INPUT AMPLIFIER. Although the output of the trigger multivibrator is always a positive step voltage, the start of the step may be initiated by either the rising (positive-going) or falling (negative-going) portion of the triggering signal. When the SLOPE switch SW15 is in the + position the base of $Q14$ is connected to the input circuit and the base of $Q24$ is connected to a bias source adjustable by means of the LEVEL control. With this configuration $Q14$-$Q24$ is an emitter-coupled amplifier and the signal at the collector of $Q24$ is in phase with the signal at the base of $Q14$. A positive-going signal at the input will therefore produce a positive-going signal at the base of $Q35$ and, as explained previously, this is the action that initiates the sweep. The sweep will then start on the positive slope (rising portion) of the triggering signal.

When the SLOPE switch is placed in the − position, the base of $Q24$ is connected to the input circuit and the base of $Q14$ is connected to the bias source. This eliminates $Q14$ from the amplifier circuit and $Q24$ becomes a collector-loaded amplifier. The output signal will then be opposite in polarity to the base signal, and so the sweep will start on the falling or negative-going portion of the triggering signal.

TRIGGERING LEVEL. The hysteresis of the trigger multivibrator ($Q35$-$Q45$) is determined by the dividers in the base circuit of each transistor. The quiescent start of the trigger-input amplifier is such that the collector voltage of $Q24$ is about in the center of the hysteresis of the multivibrator. An adjustment of the LEVEL control will vary the bias on the transistor to which it is connected. This in turn will vary the quiescent voltage at the collector of $Q24$, within the hysteresis range of the multivibrator. By adjusting the LEVEL control, the operator can select the point on the waveform at which he wishes to trigger the sweep.

AUTOMATIC TRIGGERING. The sweep can be triggered automatically, instead of manually, by turning the LEVEL control pull left to the AUTO position. In the AUTO position SW17 is opened, which alters the circuit configuration as follows: (a) $C30$ is connected into the circuit; (b) the

LEVEL control $R17$ is disconnected from circuit; and (c) all triggering input signals are *ac*-coupled through $C8$, regardless of the setting of the ac-dc switch. In the automatic (AUTO) triggering mode, the trigger multivibrator is converted from a bistable configuration to a recurrent (free-running) configuration, since the addition of $C30$ to the circuit causes the trigger multivibrator to free-run in the absence of a triggering signal. The addition of $C30$ makes the time constant such that it takes about 10 msec for the voltage at the base of $Q35$ to rise exponentially from its starting point, below cutoff, to a point where collector current can start. As $Q35$ starts conducting, its collector voltage drops. The base voltage then starts falling exponentially. When the base voltage falls below cutoff, the circuit has completed one cycle of its approximately 50-cycle output. Thus, when not receiving triggers, the sweep continues at approximately a 50-cycle rate.

Sweep Generation

TIME-BASE GENERATOR.   The positive-going pulses produced by the time-base trigger circuit are differentiated in the base circuit of $Q135$ (Figure 4-15). The sharp positive spikes produced by the differentiation process are used to start the sweep; the negative spikes are not used. The time-base generator consists of three main circuits: a bistable sweep-gating multivibrator, a Miller integrator circuit, and a hold-off circuit. Transistors $Q135$ and $Q145$ comprise the sweep-gating multivibrator. In the stable state immediately following a sweep $Q135$ is nonconducting and $Q145$ is conducting.

The essential components in the Miller circuit are the Miller transistors $Q161$ and $Q163$, the emitter follower $Q173$, the gating transistor $Q153$, the "disconnect" diode $D153$, the timing capacitor $C160$, and the timing resistor $R160$. The hold-off circuit consists mainly of the emitter follower $Q183$ and the hold-off capacitors $C180$ and $C181$. In the quiescent state the gating transistor $Q153$ is held in conduction (by the conducting multivibrator transistor $Q145$) and its emitter voltage is negative. This holds the cathode of $D153$ negative and forces the diode to conduct. In this state the low forward impedance of $Q153$ and $D153$ shunts the timing capacitor and prevents it from charging. This action also clamps the Miller circuit in such a way that the emitter-followers $Q163$ and $Q173$ conduct very little and the amplifier $Q161$ conducts heavily.

SWEEP GENERATION.   The next positive trigger to arrive at the base of $Q135$ will force the sweep-gating multivibrator into its second stable state in which $Q145$ is cut off. The rise in voltage at the collector of $Q145$ pulls up the base of $Q153$ and this stage cuts off. The rise in voltage at the emitter of $Q153$ then back biases $D153$ and the diode stops conducting.

This action unclamps the Miller circuit and permits it to seek its own voltages.

The base of $Q163$ then starts positive, since it is connected through the timing resistor to the +45-V bus (when the VARIABLE control is in the CALIB. position). The emitter of $Q163$ and the base of $Q161$ also start positive, following the base of $Q163$. The collector of $Q161$ then starts negative, carrying with it the base and emitter of $Q173$. This causes the voltage at the lower side of the timing capacitor to increase in the negative direction, which in turn pulls down the base of $Q163$ and prevents it from going positive. The gain of the Miller circuit is such that the feedback network maintains the voltage at the base of $Q163$ virtually constant (within about one-tenth of a volt).

The timing capacitor then starts charging with current through the timing resistor and the EF $Q173$. Since the voltage at the base of $Q163$ remains essentially constant, the voltage across the timing resistor, and hence the charging current through it, remains essentially constant. The timing capacitor therefore charges linearly, and the voltage at the emitter of $Q173$ increases linearly (in the negative direction). Any departure from a linear increase in the voltage at this point will produce a change in the voltage at the base of $Q163$ in a direction to correct for the error.

The linear increase in voltage at the emitter of $Q173$ is used as the sweep time base. Timing capacitor $C160$ and timing resistor $R160$ are selected by the setting of the TIME/DIV. switch SW160. The timing resistor determines the current that charges the timing capacitor. By means of the TIME/DIV. switch, both the size of the capacitor being charged and the charging current can be selected to cover a wide range of sweep speeds. The setting of the TIME/DIV. switch therefore determines the speed at which the spot moves across the CRT. The VARIABLE control $R160V$ connects the timing resistor to a voltage adjustable between +45 and +10V, which varies the sweep rate over about a 4 to 1 range. The SLOW SWP ADJ $R167$ is adjusted to regulate the current through $Q163$ so that no base current flows through $R161$ at slow sweep rates.

RETRACE AND HOLD-OFF. The distance the spot moves across the CRT is determined by the setting of the SWP LENGTH control $R176$. As the sweep voltage increases negatively at the emitter of $Q173$, there is a linear increase in voltage at the arm of $R176$ and at the base of $Q183$. This pulls down the voltage at the emitter of $Q183$ and at the base of $Q135$. When the voltage at the base of $Q135$ falls to cutoff, the sweep-gating multivibrator rapidly reverts to its original state with $Q135$ cut off and $Q145$ conducting. The voltage at the collector of $Q145$ then drops, carrying with it the base of $Q153$. This gates on $Q153$ and the diode

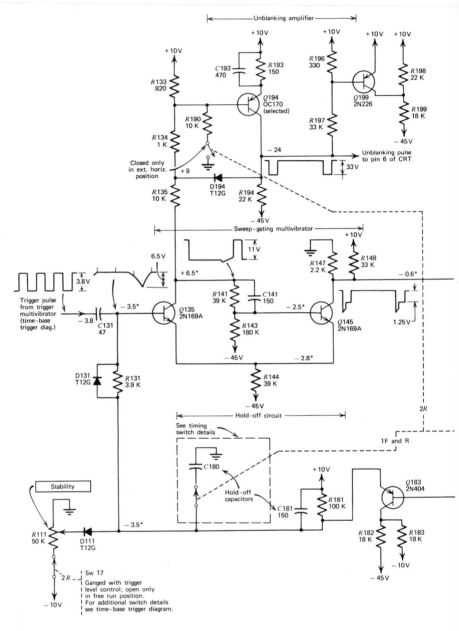

Fig. 4-15 Time-base generator circuit of the Type 321 Oscilloscope.

Waveforms and voltage readings were obtained with controls set as indicated on time-base trigger diagram; *voltage readings marked with asterisk will vary with setting of stability control

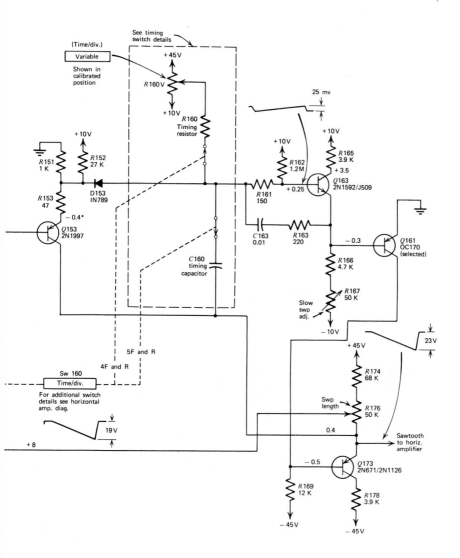

293

$D153$ and provides a discharge path for $C160$. The resistance through which $C160$ discharges ($R153$ and the forward resistance of $Q153$ and $D153$) is much less than that through which it charges ($R160$, $R178$, and the forward resistance of $Q173$). The capacitor current during discharge will therefore be much larger than during charge, and the Miller transistors will return rapidly to their quiescent state. This produces the retrace portion of the sweep sawtooth, during which time the CRT beam returns rapidly to its starting point.

The hold-off circuit prevents the time-base generator from being triggered during the retrace interval. In addition, the hold-off allows a finite time for the sweep circuits to regain a state of equilibrium after the completion of a sweep. During the trace portion of the sweep sawtooth, the hold-off capacitors $C180$-$C181$ charge through $Q183$ as a result of the drop in voltage at the emitter of $Q183$. This pulls the base of $Q135$ negative until $Q135$ cuts off. As explained previously, this is the action that initiates the retrace.

At the start of the retrace $C180$ and $C181$ start discharging through $R181$. The time constant of this circuit is such that during the retrace, and for a short period after the completion of the retrace, the base of $Q135$ is held far enough below cutoff such that positive triggers cannot switch the sweep-gating multivibrator. When the hold-off capacitors have discharged to a predetermined voltage (established by the setting of the STABILITY control), the effect of the hold-off is removed. This returns the sweep-gating multivibrator to its quiescent state, in which it can be triggered by the next positive trigger to arrive at the base of $Q135$.

STABILITY CONTROL. The STABILITY control $R111$ regulates the dc level at the base of $Q135$ within the hysteresis of the sweep-gating multivibrator. When this control is properly adjusted, the base of $Q135$ is held just negative enough that $Q135$ is back-biased and nonconducting; this prevents the circuit from free running. The base voltage must be sufficiently close to cutoff, however, that positive triggers can pull $Q135$ out of cutoff and force the multivibrator into its other state to initiate the sweep.

During the trace portion of the sweep sawtooth, when the hold-off capacitors are charging, the emitter of $Q183$ is forced negative. When the emitter of $Q183$ is more negative than the arm of the STABILITY control, the diode $D111$ is back-biased; this disconnects the STABILITY control from the multivibrator circuit.

During the retrace portion of the sweep sawtooth the hold-off capacitors discharge. When the voltage at the emitter of $Q183$ rises to the voltage at the arm of the STABILITY control the diode $D111$ conducts and clamps the hold-off circuit at this voltage. With the base of $Q135$ clamped in this manner, a sweep can only be produced when a positive trigger pulls Q 135 out of cutoff

However, should a free-running trace be desired, the (TRIGGERING) LEVEL control can be turned full right to the FREE RUN position. This opens switch $SW17$ and forces the arm of the STABILITY control to ground potential.

UNBLANKING. The CRT in Type 321 contains a second set of horizontal deflection plates (pins 6 and 10; see Figure 4-17). In the interval between sweeps, pin 6 rests at about -20 V; the CRT beam is therefore deflected off the screen and is not visible.

When a positive trigger switches the sweep-gating multivibrator to start a sweep, the negative gate at the collector of $Q135$ is coupled to the base of $Q194$. This results in a 30-V positive gate at the collector of $Q194$, which in turn is fed to pin 6 of the CRT. The 30-V positive gate, whose start and duration are coincident with the start and duration of the sweep, pulls pin 6 of the CRT up to $+10$ V, the same as pin 10. This deflects the CRT beam back into the range of visibility for the trace portion of the sweep sawtooth.

Transistor $Q199$ is connected as a load stabilizer. $Q199$ conducts when $Q194$ is nonconducting. When $Q194$ conducts during sweep time, $Q199$ is nonconducting. This circuit prevents the switching of $Q194$ from changing the load on the power supply and therefore prevents cross talk to the vertical amplifier.

Horizontal Amplifier

Figure 4-16 shows the horizontal amplifier circuit. It consists of an emitter-follower input stage $Q313$ and an emitter-coupled paraphase amplifier $Q334$-$Q344$. For all sweep-time settings of the TIME/DIV. switch, the negative-going sweep sawtooth produced by the Miller circuit is coupled to the input EF $Q313$ via the frequency-compensated voltage divider $R311$-$R312$. In the EXT. HORIZ. INPUT setting of the switch, $Q313$ receives its signal from the EXT. HORIZ. INPUT connector. This setting of the switch also produces the following results: (a) the time-base generator is rendered inoperative and (b) the collector of $Q194$ (unblanking circuit) is clamped at about $+10$ V, thus removing the unblanking potential at pin 6 of the CRT.

Emitter follower $Q323$ balances the horizontal amplifier for dc potentials. This stage also couples the positioning voltage from the HORIZONTAL POSITION control $R321$ to the amplifier circuit.

$Q334$ and $Q344$ are connected as an emitter-coupled paraphase ampli-

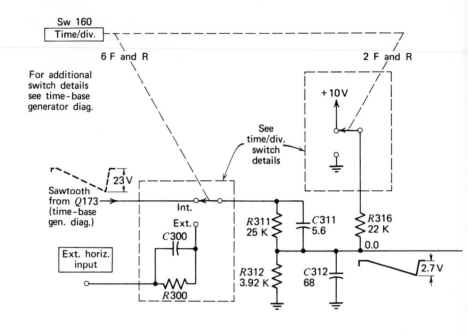

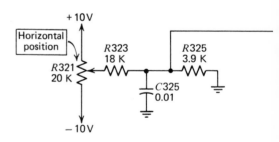

Fig. 4-16 Horizontal amplifier circuit of the Type 321 Oscilloscope.

Waveforms and voltage readings were obtained with controls set as follows:
Waveforms
    Use control settings for waveforms on time-base trigger diagram.

296

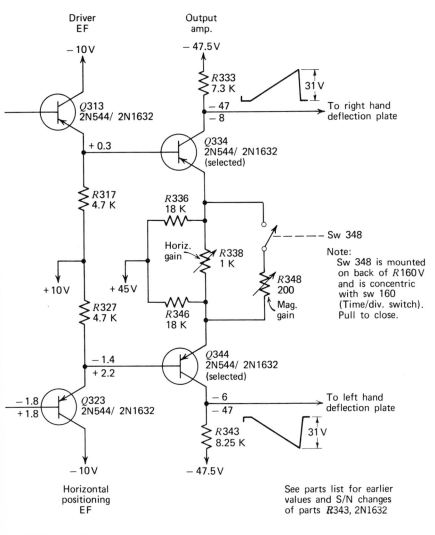

Driver EF

Output amp.

Q313 2N544/ 2N1632

Q334 2N544/ 2N1632 (selected)

R333 7.3 K

31 V

To right hand deflection plate

− 47
− 8

+ 0.3

R317 4.7 K

+ 10 V

+ 45 V

R336 18 K

Horiz. gain

R338 1 K

R348 200

Mag. gain

Sw 348

Note:
Sw 348 is mounted on back of R160V and is concentric with sw 160 (Time/div. switch). Pull to close.

R327 4.7 K

R346 18 K

− 1.4
+ 2.2

Q344 2N544/ 2N1632 (selected)

Q323 2N544/ 2N1632

− 1.8
+ 1.8

− 6
− 47

To left hand deflection plate

31 V

R343 8.25 K

− 10 V

− 47.5 V

− 10 V

− 47.5 V

Horizontal positioning EF

See parts list for earlier values and S/N changes of parts R343, 2N1632

Voltage readings
Time/div. . . . . ext. horiz. input.
Horizontal position . . . . . . full left (upper readings)
full right (lower readings)

297

fier to provide the push-pull drive for the horizontal deflection plates of the CRT. The setting of the HORIZ GAIN control $R338$ determines the emitter degeneration and thus sets the gain of the stage. A second gain control $R348$ (MAG GAIN) is connected across $R338$ when the VARIABLE timing control is pulled out to close SW348. This action decreases the degeneration and increases the gain 5 times to provide $5X$ sweep magnification.

Cathode-ray-tube Circuit

The CRT circuit is shown in Figure 4-17. The INTENSITY control $R844$, part of a divider connected between -720 volts and ground, varies the CRT grid voltage to regulate the beam current. The FOCUS control $R842$ varies the voltage at the focusing anode to set the second crossover point right at the CRT screen. The ASTIGMATISM control $R864$ varies the voltage at the astigmatism anode to focus the spot in both dimensions simultaneously. The GEOM ADJ $R861$ varies the field the beam encounters as it emerges from the deflection system to control the linearity at the extremes of deflection.

A pentupler, starting with the voltage at terminal 16 of $T701$ (Figure 4-19), builds up a potential of 3350 V for the postdeflection accelerator in the CRT. This provides an accelerating potential of approximately 4 kV, since the cathode voltage is about -670 V. Specifications of the CRT are as follows:

1. Type: special Tektronix-manufactured $T321$; 3-in flat face, postdeflection accelerator; low heater power.
2. Accelerating potential: 4 kV.
3. Z-axis modulation: external terminal permits RC coupling to grid.
4. Unblanking: deflection unblanking.
5. Phosphor: Type P2 normally furnished.

Power Supply

REGULATOR CIRCUIT. The regulator circuit is shown in Figure 4-18. Its function is to provide a regulated 10-V dc for the converter-type power supply. The regulator is designed to operate from a self-contained battery pack, from an external 11.5 to 35-V dc source, or from either a 117 or a 234-V rms, with a 50 to 800-cycle ac line.

The operation of the power-input circuit is unique in that if all three sources of power (ac, external dc, or battery) are connected to the instrument, the one providing the highest voltage will automatically be connected to the regulator. This is accomplished by the action of the "disconnect" diodes $D620$, $D621$, and $D622$.

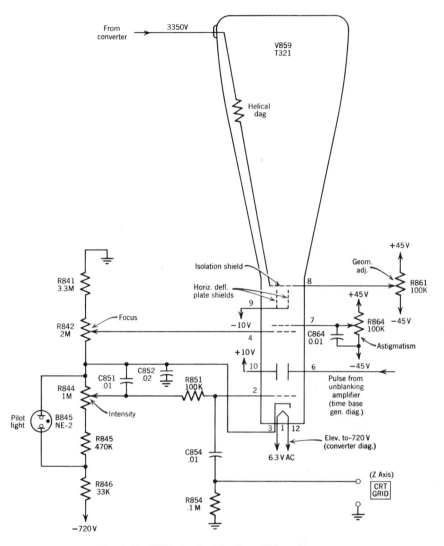

Fig. 4-17 CRT circuit of the Type 321 oscilloscope.

The Zener diode *D*629 (350 up, 152 to 016, Tektronix) provides a constant dc voltage of about +6.2 V with respect to the common negative bus, at the base of *Q*624, one-half of a difference amplifier. The base voltage for the other half of the difference amplifier *Q*634 is obtained from a divider *R*650, *R*651, and *R*652. The 10-V ADJ control *R*651

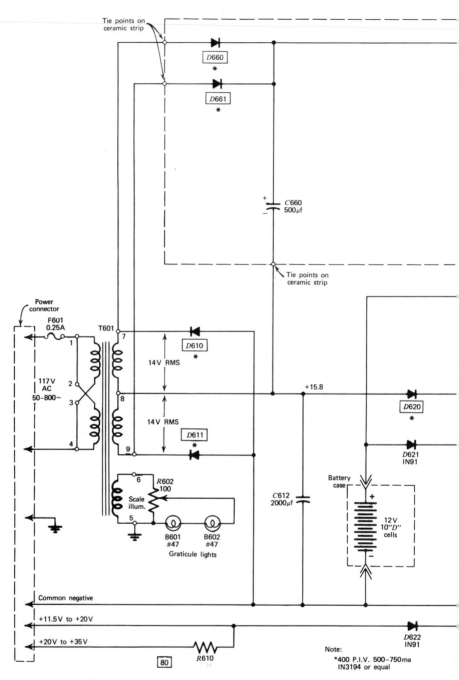

Fig. 4-18   Regulator circuit of the Type 321 Oscilloscope.

300

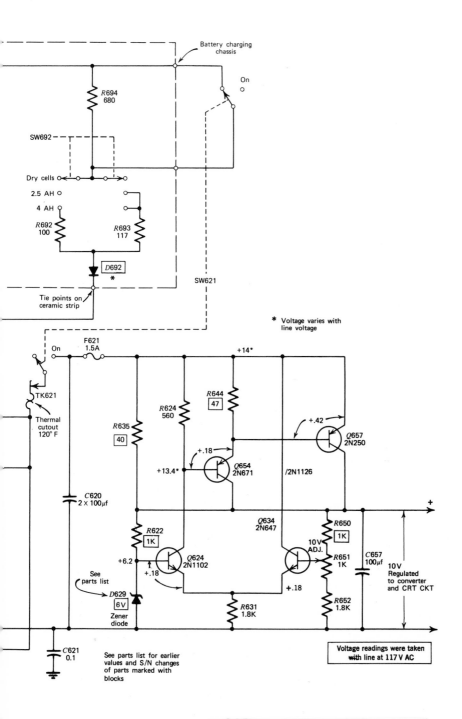

Battery charging chassis

On

R694
680

SW692

Dry cells

2.5 AH

4 AH

R692
100

R693
117

D692
*

Tie points on ceramic strip

SW621

* Voltage varies with line voltage

On

F621
1.5A

+14*

R644
47

TK621

Thermal cutout 120° F

R624
560

R635
40

Q657
2N250
+.42

R622
1K

+.18

Q654
2N671
+13.4*

/2N1126

C620
2 × 100μf

Q634
2N647

R650
1K

+

+6.2

Q624
2N1102

10V
ADJ.

R651
1K

C657
100μf

10V
Regulated
to converter
and CRT CKT

+.18

See parts list

D629
6V
Zener
diode

R631
1.8K

+.18

R652
1.8K

C621
0.1

See parts list for earlier
values and S/N changes
of parts marked with
blocks

Voltage readings were taken
with line at 117 V AC

301

determines the percentage of total voltage that appears at the base of Q634 and thus determines the total voltage across the divider. This control is adjusted so that the output is exactly 10 V.

The regulation of the output, in the presence of line-voltage or load changes, is accomplished by varying the impedance of Q657 in a direction that compensates for the change. For example, assume the output voltage tends to decrease. This will lower the base voltage of Q634 and alter the current distribution through the difference amplifier. That is, the current through Q634 will decrease and the current through Q624 will increase. The resultant drop in voltage at the collector of Q624 will pull down the base and emitter of Q654. This action drives the base of Q657 toward its collector, which increases the current through Q657 and lowers its impedance. The voltage across C620 is equal to the output voltage plus the drop across Q657. The decrease in impedance of Q657 lowers its emitter-to-collector voltage, which causes the output voltage to increase to its proper value. If the output voltage tends to increase, just the opposite will occur.

BATTERY CHARGER. When type 321 is connected to a source of ac power and the POWER switch is turned off, the rectified output T601 is sufficient to cause a back current through the 12-volt battery, resulting in a charging of the cells.

CONVERTER. The converter circuit is shown in Figure 4-19. The regulated 10-V output of the regulator circuit is applied to the converter transistors Q700-Q710. Both transistors are biased in the forward direction when power is applied, but because of slight difference in characteristics one will start conducting before the other. Current flowing in the collector circuit of the conducting transistor will then induce a voltage into the base windings (terminal 5, 6, and 7) of transformer T701. The polarity of the base voltages induced will be such that the conducting transistor will conduct more, and the nonconducting transistor will be driven into cutoff.

The buildup of current in the conducting transistor will continue until the transformer saturates. At saturation the induced voltage in the base windings will start decaying and the conducting transistor will accordingly conduct less. The collapsing field will then induce voltages of the opposite polarity in the base windings. This will drive the transistor that had been conducting into cutoff, and turn on the transistor that had previously been cut off. The circuit will then produce a secondary voltage of the opposite polarity. The repetition rate of the circuit is about 2 kc.

The transistor current, flowing in the primary circuit of T701, is trapezoidal in shape; the secondary voltage is therefore a square wave. A square wave of approximately 100 V, peak-to-peak, is coupled from terminal 10 of $T701$ to the calibrator circuit (Figure 4-12).

4-6 TYPICAL SCALER-INTEGRATED CIRCUIT

Type SN7490N Decade Counter

The decade counter (without feedback) was discussed in Sec. 1-19; its transistorized circuit is shown in Figure 1-131. The decade-counter integrated circuits are available commercially. The primary advantages of the integrated circuits over the transistorized circuits are increased reliability, higher speed, reduced physical size, and lower cost.

Figure 4-20 shows the Texas Instruments SN7490N decade counter with its truth tables. It is a high-speed monolithic decade counter consisting of four dual-rank master-slave flip-flops internally interconnected to provide a divide-by-two counter and a divide-by-five counter. Gated directed reset lines are provided to inhibit count inputs and return all outputs to a logical zero or to a binary coded decimal (BCD) count of 9. As the output from flip-flop A is not internally connected to the succeeding stages, the count may be separated into three independent count modes:

1. When used as a BCD decade counter, the BD input must be externally connected to the A output. The A input receives the incoming count, and a count sequence is obtained in accordance with the BCD count sequence truth table shown in Figure 4-20. In addition to a conventional zero reset, inputs are provided to reset a BCD 9 count for nine's compliment decimal applications.

2. If a symmetrical divide-by-ten count is desired for frequency synthesizers or other applications resulting division of a binary count by a power of 10, the D output must be externally connected to the A input. The input count is then applied at the BD input and a divide-by-ten square wave is obtained at output A.

3. For operation as a divide-by-two counter and a divide-by-five counter, no external interconnections are required. Flip-flop A is used as a binary element for the divide-by-two function. The BD input is used to obtain binary divide-by-five operation at the B, C, and D outputs. In this mode the two counters operate independently; however, all four flip-flops are reset simultaneously.

Characteristics, circuits, and parameter measurement information of the SN7490N are given in Appendix 5. This counter is completely compatible with Series 74 and Series 74 930 transistor transistor logic (TTL), and Series 15830 diode transistor logic (DTL) families. Average power dissipation is 160 mW.

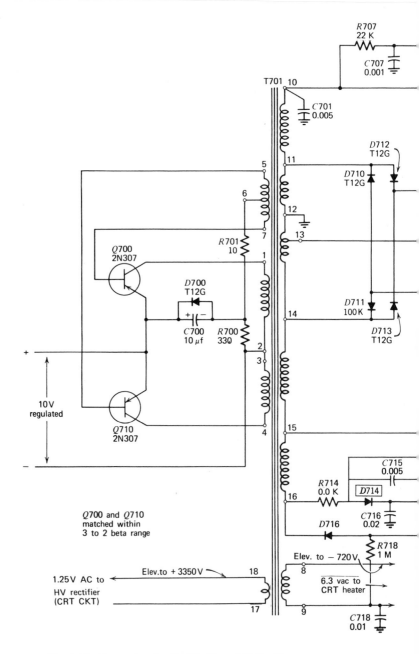

Fig. 4-19   Converter circuit of the Type 321 Oscilloscope.

304

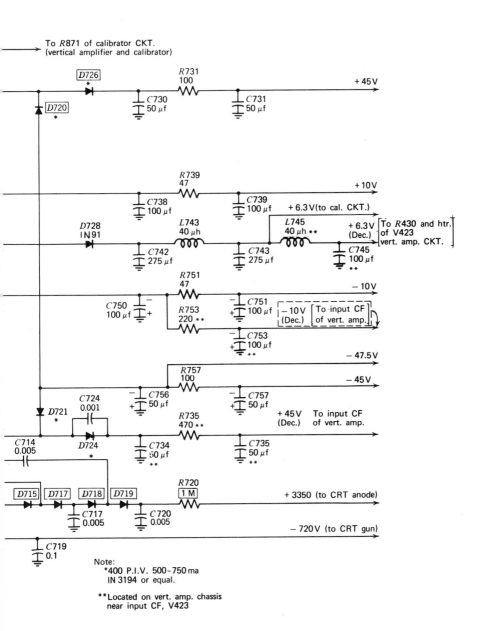

To R871 of calibrator CKT.
(vertical amplifier and calibrator)

D726
D720

R731
100
C730 50 μf
C731 50 μf
+ 45 V

R739
47
C738 100 μf
C739 100 μf
+ 10 V

+ 6.3 V (to cal. CKT.)

D728
IN91

L743
40 μh
C742 275 μf

L745
40 μh **
C743 275 μf

+ 6.3 V
(Dec.)

To R430 and htr.
of V423
vert. amp. CKT.

C745 100 μf **

R751
47
C750 100 μf

R753 220 **
C751 100 μf
− 10 V

− 10 V
(Dec.)
To input CF
of vert. amp.

C753 100 μf **

− 47.5 V

R757 100
C756 50 μf
C757 50 μf
− 45 V

C724 0.001
D721

R735 470 **
+ 45 V
(Dec.)
To input CF
of vert. amp.

C714 0.005
D724
C734 50 μf **
C735 50 μf **

D715 D717 D718 D719
R720
1 M
+ 3350 (to CRT anode)

C717 0.005
C720 0.005

− 720 V (to CRT gun)

C719 0.1

Note:
*400 P.I.V. 500-750 ma
IN 3194 or equal.

**Located on vert. amp. chassis
near input CF, V423

305

Truth tables

BCD count sequence
(See note 1)

Count	Output			
	D	C	B	A
0	0	0	0	0
1	0	0	0	1
2	0	0	1	0
3	0	0	1	1
4	0	1	0	0
5	0	1	0	1
6	0	1	1	0
7	0	1	1	1
8	1	0	0	0
9	1	0	0	1

Reset/count (see note 2)

$R_{0(1)}$	$R_{0(2)}$	$R_{9(1)}$	$R_{9(2)}$	Output D C B A
1	1	0	X	0 0 0 0
1	1	X	0	0 0 0 0
X	X	1	1	1 0 0 1
X	0	X	0	Count
0	X	0	X	Count
0	X	X	0	Count
X	0	0	X	Count

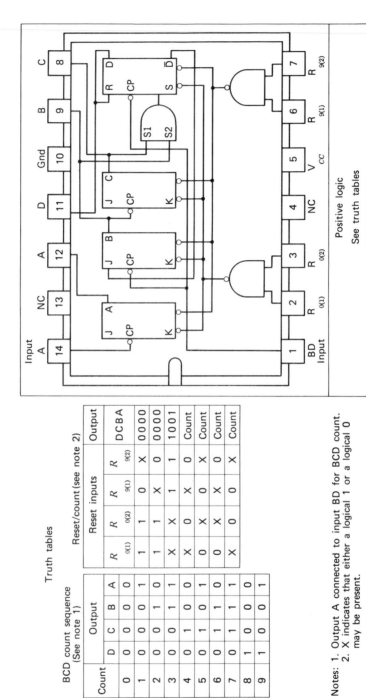

Positive logic
See truth tables

Fig. 4-20 The Texas Instruments Type SN7490N decade counter with its truth tables.

Notes: 1. Output A connected to input BD for BCD count.
2. X indicates that either a logical 1 or a logical 0 may be present.

306

Type SN7441N BCD-to-Decimal Decoder/Driver

Figure 4-21 shows the Texas Instruments type SN7441N logic with its truth and decoding table. This complex-function integrated circuit is a monolithic BCD-to-decimal decoder with output transistors for directly driving gas-filled readout tubes or miniature dc components. The decoder consists of familiar TTL gate circuits that receive and decode BCD-coded inputs and select one of the 10 decimal output driver transistors. The BCD inputs are fully compatible with Series 74 logic out-

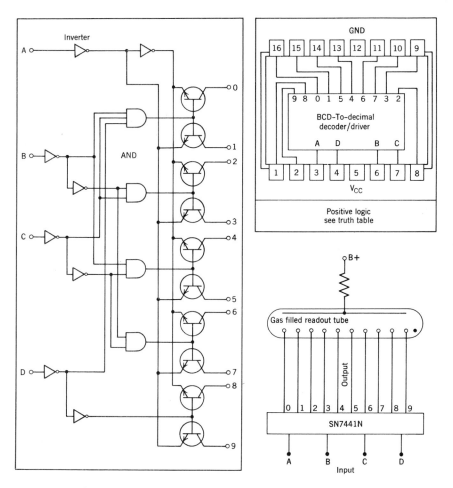

Fig. 4-21   Type SN7441N logic with its truth and decoding table.

puts; and, in addition, physical placement of these inputs is coincidental with the BCD outputs of the SN7490N decade counter. The 10 high-breakdown *n-p-n* transistors have a maximum leakage current of 200 μA or 55V over the operating temperature range. Decoding and dc switching for components such as miniature lamps and relays may be performed by the SN7441 within the ranges specified for the electrical characteristics. A typical application shows the SN7441N used as a gas-filled display-indicator driver.

Electric characteristics, parameter-measurement information, and typical application of the Texas Type SN7441N are given in Appendix 6.

ANL Model S194 Six-Decade Scaler with Binary-to-Decimal Decoder

INTRODUCTION.   The ANL Model S194 shown in Figure 4-22 is a six-decade scaler made up of integrated circuits. It has a 10-MHz counting rate and a 50-nsec double-pulse resolution. Either an electrical reset register of 4-digits–10-impulses/sec or a mechanical reset register of 6-digit–25-impulses/sec is used. All the indicators are low-failure neon pressfit cartridge lamps. Individual POSITIVE or NEGATIVE INPUTS are available on the front and back panels. Inputs are direct coupled (DC.) and have an input impedance of approximately 1000 Ω. Input sensitivity levels are approximately 1 V with fast rise times, and they decrease with increasing rise times. A nominal 3-V input requires a rise time of 1 μsec or faster. With the COUNT-REMOTE switch in REMOTE position any external grounding pulse will start scaler. To gate using the REMOTE jack, the gating pulse should nominally be a +3 V(off) and a +1.3 to 0 V(on). There is no maximum gating width; however, there is a minimum gating width of 200 nsec. Power requirements are +12 V, 200 mA; −12 V, 10 mA; and −24 V, 10 mA (10^3 scaler with electrical reset register).

SCALER CIRCUITRY.   The board actually contains six decades, but only four are shown in Figure 4-22. Each decade consists of a SN7490N decade counter and a SN7441N BCD-to-decimal decoder. The input pulse to the scaler is obtained from the control circuit shown in Figure 4-24, which shapes the incoming pulse and also allows for external control (gating) of the incoming signal. The input pulse triggers the first decade from which the BCD number is applied to the appropriate decoder driver so that the output count is shown by the indicator lamp. The final decade is followed by a register, either mechanical or electrical type, as shown in Figure 4-23.

Resetting of the scaler is accomplished using the NAND gate R_0 as a

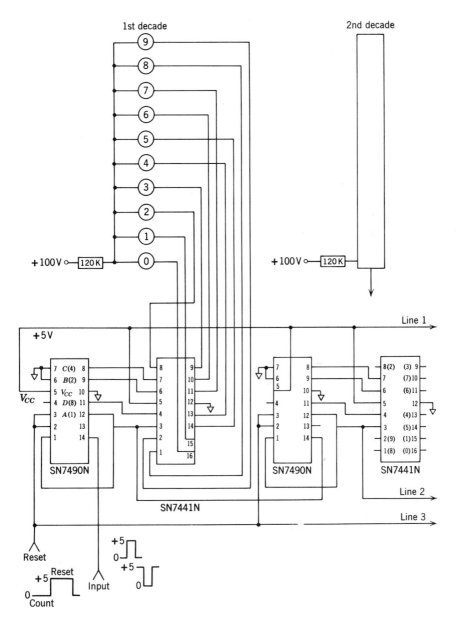

Fig. 4-22*a* First two decades of the ANL Model S194 Six Decade Scaler. For the complete circuit, this section is connected to that shown in Figure 4-22*b* by Lines 1, 2 & 3.

309

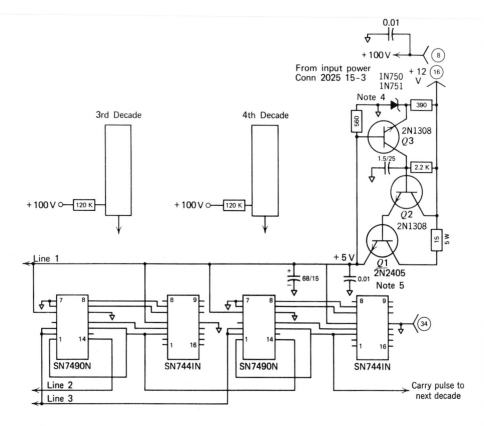

Fig. 4-22*b*   3rd and 4th decades of the ANL Model S194 Six Decade Scaler, 1968.

Notes
1. Board contains six decades but only four are shown on drawing; as many decades as desired can be used
2. Designed to be used in double nim module
3. PC board #3PC859
4. Select Zener for 5 V @ 20 mA
5. Use Wakefield type NF207 heatsink on 2N2405
6. +5V regulation ≈0.5% 10-400 mA

simple inverter, which is incorporated in the SN7490. A low input (zero state) produces a high output from the gate which turns the scaler on. A high input (1 state) to the gate produces a low output which effectively grounds the binaries and produces a reset to zero condition.

SCALER CONTROL CIRCUIT.   Figure 4-24 shows the fast shaping circuit for the scaler control. It will accept inputs of either a negative or positive polarity. The inputs are direct coupled to a limit current which keeps any large input pulses from overloading the input transistors ($Q1$ and $Q2$). Both $Q1$ and $Q2$ are very fast (300 MHz or faster) transistors and are driven to saturation by the incoming pulse. The inverted output pulses from $Q1$ and $Q2$ are collector coupled to two NAND gates of a fourfold NAND-gated integrated circuit (MC846).[1] The positive input section

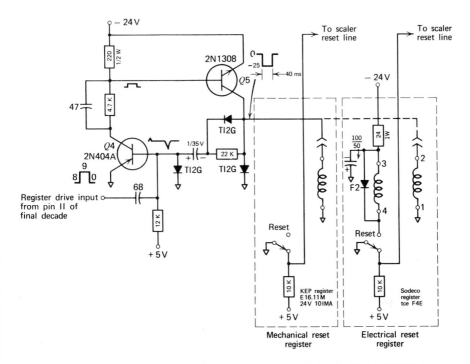

Fig. 4-23   Either mechanical or electrical reset register, used with the Model S194 Scaler.

Notes for Figs. 4-23 and 24
1. All resistors ¼W, 5% Cazson unless otherwise specified
2. Rise time limit: 200 ns @ 5V
3. PC board #3PC872
4. Mechanical reset register connected in circuit
5. Two NAND gates of MC846 used as inverters
6. Two SG5000 diodes comprise an AND gate

[1]MC846F consists of four 2-input NAND/NOR logic circuits. This gate device may be used as four inverter circuits or as two bistable circuits by cross-coupling two dual 2-input gates. Among other applications, this device can be used as a dual 2-input noninverting gate.

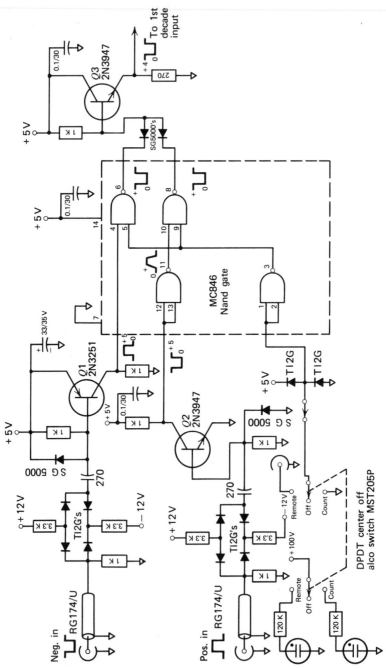

312

Fig. 4-24   ANL Model CD-617 Scaler Control, used with the Model S194 Scaler.

uses one section of the fourfold NAND gate as a simple inverter. The NAND function of the gates is supplied through the remote input jack. Any signal approaching ground through the remote input will activate the NAND gate which will pass the signal from either the positive or negative input (whichever has an input signal at the time). The signal is then fed to an emitter follower which matches the control output to the input impedance of the first decade of the 10-MHz scaler.

REFERENCES

Ello, J. G., ANL, Radiological Survey Meters, *Electron. World,* (January, 1966).

Technical Manual, *Geiger Counter Model* E-120/E-120G, Eberline Instrument Corporation, Santa Fe, New Mexico.

Instruction Manual for Model 440 *Low Energy Survey Meter, Form* 1082C-64, *Part* No. 440-2, The Victoreen Instrument Company, 5806 Hough Avenue, Cleveland, 1964.

Technical Manual, *Gas Proportional Alpha Counter Model* PAC-4G, Eberline Instrument Corporation, Santa Fe, New Mexico.

Instruction Manual for Type 321 Oscilloscope, Tektronix Manufacturers of Cathode-Ray Oscilloscopes, S. W. Millikan Way, P.O. Box 500, Beaverton, Oregon, 1961.

Cambion Integrated Circuit Logic Manual, Cambridge Thermionic Corporation, 445 Concord Avenue, Cambridge, Mass., 1967.

TTL Integrated Circuits from Texas Instruments, Canton, Mass., 02021, P. O. Box 121, 1967.

APPENDIX 1

NATURAL (NAPIERIAN) LOGARITHMS

The natural logarithm of a number is the index of the power to which the base e (2.7182818) must be raised in order to equal the number. For example,

$$\log_e 4.12 = \ln 4.12 = 1.4159$$

The table gives the natural logarithms of numbers from 1.00 to 9.99 directly, and permits finding logarithms of numbers outside that range by the addition or subtraction of the natural logarithms of powers of 10. For example,

$$\ln 679. = \ln 6.79 + \ln 10^2 = 1.9155 + 4.6052 = 6.5207,$$
$$\ln 0.0679 = \ln 6.79 - \ln 10^2 = 1.9155 - 4.6052 = -2.6897.$$

Natural Logarithms of 10^k

$\ln 10 = 2.302585$	$\ln 10^4 = 9.210340$	$\ln 10^7 = 16.118096$
$\ln 10^2 = 4.605170$	$\ln 10^5 = 11.512925$	$\ln 10^8 = 18.420681$
$\ln 10^3 = 6.907755$	$\ln 10^6 = 13.815511$	$\ln 10^9 = 20.723266$

To obtain the common logarithm, the natural logarithm is multiplied by $\log_{10} e$, which is 0.434294, or $\log_{10} N = 0.434294 \ln N$.

315

Natural Logarithms

N	0	1	2	3	4	5	6	7	8	9
1.0	0.0000	0.0100	0.0198	0.0296	0.0392	0.0488	0.0583	0.0677	0.0770	0.0862
1.1	0.0953	0.1044	0.1133	0.1222	0.1310	0.1398	0.1484	0.1570	0.1655	0.1740
1.2	0.1823	0.1906	0.1989	0.2070	0.2151	0.2231	0.2311	0.2390	0.2469	0.2546
1.3	0.2624	0.2700	0.2776	0.2852	0.2927	0.3001	0.3075	0.3148	0.3221	0.3293
1.4	0.3365	0.3436	0.3507	0.3577	0.3646	0.3716	0.3784	0.3853	0.3920	0.3988
1.5	0.4055	0.4121	0.4187	0.4253	0.4313	0.4383	0.4447	0.4511	0.4574	0.4637
1.6	0.4700	0.4762	0.4824	0.4886	0.4947	0.5008	0.5068	0.5128	0.5188	0.5247
1.7	0.5306	0.5365	0.5423	0.5481	0.5539	0.5596	0.5653	0.5710	0.5766	0.5822
1.8	0.5878	0.5933	0.5988	0.6043	0.6098	0.6152	0.6206	0.6259	0.6313	0.6366
1.9	0.6419	0.6471	0.6523	0.6575	0.6627	0.6678	0.6729	0.6780	0.6831	0.6881
2.0	0.6931	0.6981	0.7031	0.7080	0.7129	0.7178	0.7227	0.7275	0.7324	0.7372
2.1	0.7419	0.7467	0.7514	0.7561	0.7608	0.7655	0.7701	0.7747	0.7793	0.7839
2.2	0.7885	0.7930	0.7975	0.8020	0.8065	0.8109	0.8154	0.8198	0.8242	0.8286
2.3	0.8329	0.8372	0.8416	0.8459	0.8502	0.8544	0.8587	0.8629	0.8671	0.8713
2.4	0.8755	0.8796	0.8838	0.8879	0.8920	0.8961	0.9002	0.9042	0.9083	0.9123
2.5	0.9163	0.9203	0.9243	0.9282	0.9322	0.9361	0.9400	0.9439	0.9478	0.9517
2.6	0.9555	0.9594	0.9632	0.9670	0.9708	0.9746	0.9783	0.9821	0.9858	0.9895
2.7	0.9933	0.9969	1.0006	1.0043	1.0080	1.0116	1.0152	1.0188	1.0225	1.0260
2.8	1.0296	1.0332	1.0367	1.0403	1.0438	1.0473	1.0508	1.0543	1.0578	1.0613
2.9	1.0647	1.0682	1.0716	1.0750	1.0784	1.0818	1.0852	1.0886	1.0919	1.0953
3.0	1.0986	1.1019	1.1053	1.1086	1.1119	1.1151	1.1184	1.1217	1.1249	1.1282
3.1	1.1314	1.1346	1.1378	1.1410	1.1442	1.1474	1.1506	1.1537	1.1569	1.1600
3.2	1.1632	1.1663	1.1694	1.1725	1.1756	1.1787	1.1817	1.1848	1.1878	1.1909
3.3	1.1939	1.1969	1.2000	1.2030	1.2060	1.2090	1.2119	1.2149	1.2179	1.2208
3.4	1.2238	1.2267	1.2296	1.2326	1.2355	1.2384	1.2413	1.2442	1.2470	1.2499
3.5	1.2528	1.2556	1.2585	1.2613	1.2641	1.2669	1.2698	1.2726	1.2754	1.2782
3.6	1.2809	1.2837	1.2865	1.2892	1.2920	1.2947	1.2975	1.3002	1.3029	1.3056
3.7	1.3083	1.3110	1.3137	1.3164	1.3191	1.3218	1.3244	1.3271	1.3297	1.3324
3.8	1.3350	1.3376	1.3403	1.3429	1.3455	1.3481	1.3507	1.3533	1.3558	1.3584
3.9	1.3610	1.3635	1.3661	1.3686	1.3712	1.3737	1.3762	1.3788	1.3813	1.3838
4.0	1.3863	1.3888	1.3913	1.3938	1.3962	1.3987	1.4012	1.4036	1.4061	1.4085
4.1	1.4110	1.4134	1.4159	1.4183	1.4207	1.4231	1.4255	1.4279	1.4303	1.4327
4.2	1.4351	1.4375	1.4398	1.4422	1.4446	1.4469	1.4493	1.4516	1.4540	1.4563
4.3	1.4586	1.4609	1.4633	1.4656	1.4679	1.4702	1.4725	1.4748	1.4770	1.4793
4.4	1.4816	1.4839	1.4861	1.4884	1.4907	1.4929	1.4951	1.4974	1.4996	1.5019
4.5	1.5041	1.5063	1.5085	1.5107	1.5129	1.5151	1.5173	1.5195	1.5217	1.5239
4.6	1.5261	1.5282	1.5304	1.5326	1.5347	1.5369	1.5390	1.5412	1.5433	1.5454
4.7	1.5476	1.5497	1.5518	1.5539	1.5560	1.5581	1.5602	1.5623	1.5644	1.5665
4.8	1.5686	1.5707	1.5728	1.5748	1.5769	1.5790	1.5810	1.5831	1.5851	1.5872
4.9	1.5892	1.5913	1.5933	1.5953	1.5974	1.5994	1.6014	1.6034	1.6054	1.6074
5.0	1.6094	1.6114	1.6134	1.6154	1.6174	1.6194	1.6214	1.6233	1.6253	1.6273
5.1	1.6292	1.6312	1.6332	1.6351	1.6371	1.6390	1.6409	1.6429	1.6448	1.6467
5.2	1.6487	1.6506	1.6525	1.6544	1.6563	1.6582	1.6601	1.6620	1.6639	1.6658
5.3	1.6677	1.6696	1.6715	1.6734	1.6752	1.6771	1.6790	1.6808	1.6827	1.6845

Natural Logarithms (continued)

N	0	1	2	3	4	5	6	7	8	9
5.4	1.6864	1.6882	1.6901	1.6919	1.6938	1.6956	1.6974	1.6993	1.7011	1.7029
5.5	1.7047	1.7066	1.7084	1.7102	1.7120	1.7138	1.7156	1.7174	1.7192	1.7210
5.6	1.7228	1.7246	1.7263	1.7281	1.7299	1.7317	1.7334	1.7352	1.7370	1.7387
5.7	1.7405	1.7422	1.7440	1.7457	1.7475	1.7492	1.7509	1.7527	1.7544	1.7561
5.8	1.7579	1.7596	1.7613	1.7630	1.7647	1.7664	1.7681	1.7699	1.7716	1.7733
5.9	1.7750	1.7766	1.7783	1.7800	1.7817	1.7834	1.7851	1.7867	1.7884	1.7901
6.0	1.7918	1.7934	1.7951	1.7967	1.7984	1.8001	1.8017	1.8034	1.8050	1.8066
6.1	1.8083	1.8099	1.8116	1.8132	1.8148	1.8165	1.8181	1.8197	1.8213	1.8229
6.2	1.8245	1.8262	1.8278	1.8294	1.8310	1.8326	1.8342	1.8358	1.8374	1.8390
6.3	1.8405	1.8421	1.8437	1.8453	1.8469	1.8485	1.8500	1.8516	1.8532	1.8547
6.4	1.8563	1.8579	1.8594	1.8610	1.8625	1.8641	1.8656	1.8672	1.8687	1.8703
6.5	1.8718	1.8733	1.8749	1.8764	1.8779	1.8795	1.8810	1.8825	1.8840	1.8856
6.6	1.8871	1.8886	1.8901	1.8916	1.8931	1.8946	1.8961	1.8976	1.8991	1.9006
6.7	1.9021	1.9036	1.9051	1.9066	1.9081	1.9095	1.9110	1.9125	1.9140	1.9155
6.8	1.9169	1.9184	1.9199	1.9213	1.9228	1.9242	1.9257	1.9272	1.9286	1.9301
6.9	1.9315	1.9330	1.9344	1.9359	1.9373	1.9387	1.9402	1.9416	1.9430	1.9445
7.0	1.9459	1.9478	1.9488	1.9502	1.9516	1.9530	1.9544	1.9559	1.9573	1.9587
7.1	1.9601	1.9615	1.9629	1.9643	1.9657	1.9671	1.9685	1.9699	1.9713	1.9727
7.2	1.9741	1.9755	1.9769	1.9782	1.9796	1.9810	1.9824	1.9838	1.9851	1.9865
7.3	1.9879	1.9892	1.9906	1.9920	1.9933	1.9947	1.9961	1.9974	1.9988	2.0001
7.4	2.0015	2.0028	2.0042	2.0055	2.0069	2.0082	2.0096	2.0109	2.0122	2.0136
7.5	2.0149	2.0162	2.0176	2.0189	2.0202	2.0215	2.0229	2.0242	2.0255	2.0268
7.6	2.0281	2.0295	2.0308	2.0321	2.0334	2.0347	2.0360	2.0373	2.0386	2.0399
7.7	2.0412	2.0425	2.0438	2.0451	2.0464	2.0477	2.0490	2.0503	2.0516	2.0528
7.8	2.0541	2.0554	2.0567	2.0580	2.0592	2.0605	2.0618	2.0631	2.0643	2.0656
7.9	2.0669	2.0681	2.0694	2.0707	2.0719	2.0732	2.0744	2.0757	2.0769	2.0782
8.0	2.0794	2.0807	2.0819	2.0832	2.0844	2.0857	2.0869	2.0882	2.0894	2.0906
8.1	2.0919	2.0931	2.0943	2.0956	2.0968	2.0980	2.0992	2.1005	2.1017	2.1029
8.2	2.1041	2.1054	2.1066	2.1078	2.1090	2.1102	2.1114	2.1126	2.1138	2.1150
8.3	2.1163	2.1175	2.1187	2.1199	2.1211	2.1223	2.1235	2.1247	2.1258	2.1270
8.4	2.1282	2.1294	2.1306	2.1318	2.1330	2.1342	2.1353	2.1365	2.1377	2.1389
8.5	2.1401	2.1412	2.1424	2.1436	2.1448	2.1459	2.1471	2.1483	2.1494	2.1506
8.6	2.1518	2.1529	2.1541	2.1552	2.1564	2.1576	2.1587	2.1599	2.1610	2.1622
8.7	2.1633	2.1645	2.1656	2.1668	2.1679	2.1691	2.1702	2.1713	2.1725	2.1736
8.8	2.1748	2.1759	2.1770	2.1782	2.1793	2.1804	2.1815	2.1827	2.1838	2.1849
8.9	2.1861	2.1872	2.1883	2.1894	2.1905	2.1917	2.1928	2.1939	2.1950	2.1961
9.0	2.1972	2.1983	2.1994	2.2006	2.2017	2.2028	2.2039	2.2050	2.2061	2.2072
9.1	2.2083	2.2094	2.2105	2.2116	2.2127	2.2138	2.2148	2.2159	2.2170	2.2181
9.2	2.2192	2.2203	2.2214	2.2225	2.2235	2.2246	2.2257	2.2268	2.2279	2.2289
9.3	2.2300	2.2311	2.2322	2.2332	2.2343	2.2354	2.2364	2.2375	2.2386	2.2396
9.4	2.2407	2.2418	2.2428	2.2439	2.2450	2.2460	2.2471	2.2481	2.2492	2.2502
9.5	2.2513	2.2523	2.2534	2.2544	2.2555	2.2565	2.2576	2.2586	2.2597	2.2607
9.6	2.2618	2.2628	2.2638	2.2649	2.2659	2.2670	2.2680	2.2690	2.2701	2.2711
9.7	2.2721	2.2732	2.2742	2.2752	2.2762	2.2773	2.2783	2.2793	2.2803	2.2814
9.8	2.2824	2.2834	2.2844	2.2854	2.2865	2.2875	2.2885	2.2895	2.2905	2.2915
9.9	2.2925	2.2935	2.2946	2.2956	2.2966	2.2976	2.2986	2.2996	2.3006	2.3016

APPENDIX 2

Logarithms to Base 10

N	0	1	2	3	4	5	6	7	8	9	1	2	3	4	5	6	7	8	9
10	0000	0043	0086	0128	0170	0212	0253	0294	0334	0374	4	8	12	17	21	25	29	33	37
11	0414	0453	0492	0531	0569	0607	0645	0682	0719	0755	4	8	11	15	19	23	26	30	34
12	0792	0828	0864	0899	0934	0969	1004	1038	1072	1106	3	7	10	14	17	21	24	28	31
13	1139	1173	1206	1239	1271	1303	1335	1367	1399	1430	3	6	10	13	16	19	23	26	29
14	1461	1492	1523	1553	1584	1614	1644	1673	1703	1732	3	6	9	12	15	18	21	24	27
15	1761	1790	1818	1847	1875	1903	1931	1959	1987	2014	3	6	8	11	14	17	20	22	25
16	2041	2068	2095	2122	2148	2175	2201	2227	2253	2279	3	5	8	11	13	16	18	21	24
17	2304	2330	2355	2380	2405	2430	2455	2480	2504	2529	2	5	7	10	12	15	17	20	22
18	2553	2577	2601	2625	2648	2672	2695	2718	2742	2765	2	5	7	9	12	14	16	19	21
19	2788	2810	2833	2856	2878	2900	2923	2945	2967	2989	2	4	7	9	11	13	16	18	20
20	3010	3032	3054	3075	3096	3118	3139	3160	3181	3201	2	4	6	8	11	13	15	17	19
21	3222	3243	3263	3284	3304	3324	3345	3365	3385	3404	2	4	6	8	10	12	14	16	18
22	3424	3444	3464	3483	3502	3522	3541	3560	3579	3598	2	4	6	8	10	12	14	16	17
23	3617	3636	3655	3674	3692	3711	3729	3747	3766	3784	2	4	6	7	9	11	13	15	17
24	3802	3820	3838	3856	3874	3892	3909	3927	3945	3962	2	4	5	7	9	11	12	14	16
25	3979	3997	4014	4031	4048	4065	4082	4099	4116	4133	2	4	5	7	9	10	12	14	16
26	4150	4166	4183	4200	4216	4232	4249	4265	4281	4298	2	3	5	7	8	10	11	13	15
27	4314	4330	4346	4362	4378	4393	4409	4425	4440	4456	2	3	5	6	8	9	11	12	14
28	4472	4487	4502	4518	4533	4548	4564	4579	4594	4609	2	3	5	6	8	9	11	12	14
29	4624	4639	4654	4669	4683	4698	4713	4728	4742	4757	1	3	4	6	7	9	10	12	13
30	4771	4786	4800	4814	4829	4843	4857	4871	4886	4900	1	3	4	6	7	9	10	11	13
31	4914	4928	4942	4955	4969	4983	4997	5011	5024	5038	1	3	4	5	7	8	10	11	12
N	0	1	2	3	4	5	6	7	8	9	1	2	2	4	5	6	7	8	9

The proportional parts are stated in full for every tenth at the right-hand side. The logarithm of any number of four significant figures can be read directly by adding the proportional part corresponding to the fourth figure to the tabular number corresponding to the first three figures. There may be an error of 1 in the last place.

Logarithms to Base 10 (*continued*)

N	0	1	2	3	4	5	6	7	8	9	1	2	3	4	5	6	7	8	9
32	5051	5065	5079	5092	5105	5119	5132	5145	5159	5172	1	3	4	5	7	8	9	11	12
33	5185	5198	5211	5224	5237	5250	5263	5276	5289	5302	1	3	4	5	7	8	9	11	12
34	5315	5328	5340	5353	5366	5378	5391	5403	5416	5428	1	2	4	5	6	8	9	10	11
35	5441	5453	5465	5478	5490	5502	5514	5527	5539	5551	1	2	4	5	6	7	9	10	11
36	5563	5575	5587	5599	5611	5623	5635	5647	5658	5670	1	2	4	5	6	7	8	10	11
37	5682	5694	5705	5717	5729	5740	5752	5763	5775	5786	1	2	4	5	6	7	8	9	11
38	5798	5809	5821	5832	5843	5855	5866	5877	5888	5899	1	2	3	5	6	7	8	9	10
39	5911	5922	5933	5944	5955	5966	5977	5988	5999	6010	1	2	3	4	5	7	8	9	10
40	6021	6031	6042	6053	6064	6075	6085	6096	6107	6117	1	2	3	4	5	6	8	9	10
41	6128	6138	6149	6160	6170	6180	6191	6201	6212	6222	1	2	3	4	5	6	7	8	9
42	6232	6243	6253	6263	6274	6284	6294	6304	6314	6325	1	2	3	4	5	6	7	8	9
43	6335	6345	6355	6365	6375	6385	6395	6405	6415	6425	1	2	3	4	5	6	7	8	9
44	6435	6444	6454	6464	6474	6484	6493	6503	6513	6522	1	2	3	4	5	6	7	8	9
45	6532	6542	6551	6561	6571	6580	6590	6599	6609	6618	1	2	3	4	5	6	7	8	9
46	6628	6637	6646	6656	6665	6675	6684	6693	6702	6712	1	2	3	4	5	6	7	7	8
47	6721	6730	6739	6749	6758	6767	6776	6785	6794	6803	1	2	3	4	5	6	7	7	8
48	6812	6821	6830	6839	6848	6857	6866	6875	6884	6893	1	2	3	4	5	6	7	7	8
49	6902	6911	6920	6928	6937	6946	6955	6964	6972	6981	1	2	3	4	4	5	6	7	8
50	6990	6998	7007	7016	7024	7033	7042	7050	7059	7067	1	2	3	3	4	5	6	7	8
51	7076	7084	7093	7101	7110	7118	7126	7135	7143	7152	1	2	3	3	4	5	6	7	8
52	7160	7168	7177	7185	7193	7202	7210	7218	7226	7235	1	2	3	3	4	5	6	7	7
53	7243	7251	7259	7267	7275	7284	7292	7300	7308	7316	1	2	2	3	4	5	6	6	7
54	7324	7332	7340	7348	7356	7364	7372	7380	7388	7396	1	2	2	3	4	5	6	6	7
55	7404	7412	7419	7427	7435	7443	7451	7459	7466	7474	1	2	2	3	4	5	5	6	7
56	7482	7490	7497	7505	7513	7520	7528	7536	7543	7551	1	2	2	3	4	5	5	6	7
57	7559	7566	7574	7582	7589	7597	7604	7612	7619	7627	1	1	2	3	4	5	5	6	7
58	7634	7642	7649	7657	7664	7672	7679	7686	7694	7701	1	1	2	3	4	4	5	6	7
59	7709	7716	7723	7731	7738	7745	7752	7760	7767	7774	1	1	2	3	4	4	5	6	7
60	7782	7789	7796	7803	7810	7818	7825	7832	7839	7846	1	1	2	3	4	4	5	6	6
61	7853	7860	7868	7875	7882	7889	7896	7903	7910	7917	1	1	2	3	3	4	5	6	6
62	7924	7931	7938	7945	7952	7959	7966	7973	7980	7987	1	1	2	3	3	4	5	5	6
63	7993	8000	8007	8014	8021	8028	8035	8041	8048	8055	1	1	2	3	3	4	5	5	6
64	8062	8069	8075	8082	8089	8096	8102	8109	8116	8122	1	1	2	3	3	4	5	5	6
65	8129	8136	8142	8149	8156	8162	8169	8176	8182	8189	1	1	2	3	3	4	5	5	6
66	8195	8202	8209	8215	8222	8228	8235	8241	8248	8254	1	1	2	3	3	4	5	5	6
67	8261	8267	8274	8280	8287	8293	8299	8306	8312	8319	1	1	2	3	3	4	5	5	6
68	8325	8331	8338	8344	8351	8357	8363	8370	8376	8382	1	1	2	3	3	4	4	5	6
69	8388	8395	8401	8407	8414	8420	8426	8432	8439	8445	1	1	2	3	3	4	4	5	6
70	8451	8457	8463	8470	8476	8482	8488	8494	8500	8506	1	1	2	3	3	4	4	5	6
71	8513	8519	8525	8531	8537	8543	8549	8555	8561	8567	1	1	2	3	3	4	4	5	6
72	8573	8579	8585	8591	8597	8603	8609	8615	8621	8627	1	1	2	3	3	4	4	5	6
73	8633	8639	8645	8651	8657	8663	8669	8675	8681	8686	1	1	2	2	3	4	4	5	5
74	8692	8698	8704	8710	8716	8722	8727	8733	8739	8745	1	1	2	2	3	4	4	5	5
75	8751	8756	8762	8768	8774	8779	8785	8791	8797	8802	1	1	2	2	3	3	4	5	5
N	0	1	2	3	4	5	6	7	8	9	1	2	3	4	5	6	7	8	9

Logarithms to Base 10 (*continued*)

N	0	1	2	3	4	5	6	7	8	9	1	2	3	4	5	6	7	8	9
76	8808	8814	8820	8825	8831	8837	8842	8848	8854	8859	1	1	2	2	3	3	4	4	5
77	8865	8871	8876	8882	8887	8893	8899	8904	8910	8915	1	1	2	2	3	3	4	4	5
78	8921	8927	8932	8938	8943	8949	8954	8960	8965	8971	1	1	2	2	3	3	4	4	5
79	8976	8982	8987	8993	8998	9004	9009	9015	9020	9025	1	1	2	2	3	3	4	4	5
80	9031	9036	9042	9047	9053	9058	9063	9069	9074	9079	1	1	2	2	3	3	4	4	5
81	9085	9090	9096	9101	9106	9112	9117	9122	9128	9133	1	1	2	2	3	3	4	4	5
82	9138	9143	9149	9154	9159	9165	9170	9175	9180	9186	1	1	2	2	3	3	4	4	5
83	9191	9196	9201	9206	9212	9217	9222	9227	9232	9238	1	1	2	2	3	3	4	4	5
84	9243	9248	9253	9258	9263	9269	9274	9279	9284	9289	1	1	2	2	3	3	4	4	5
85	9294	9299	9304	9309	9315	9320	9325	9330	9335	9340	1	1	2	2	3	3	4	4	5
86	9345	9350	9355	9360	9365	9370	9375	9380	9385	9390	1	1	2	2	3	3	4	4	5
87	9395	9400	9405	9410	9415	9420	9425	9430	9435	9440	1	1	2	2	3	3	4	4	5
88	9445	9450	9455	9460	9465	9469	9474	9479	9484	9489	0	1	1	2	2	3	3	4	4
89	9494	9499	9504	9509	9513	9518	9523	9528	9533	9538	0	1	1	2	2	3	3	4	4
90	9542	9547	9552	9557	9562	9566	9571	9576	9581	9586	0	1	1	2	2	3	3	4	4
91	9590	9595	9600	9605	9609	9614	9619	9624	9628	9633	0	1	1	2	2	3	3	4	4
92	9638	9643	9647	9652	9657	9661	9666	9671	9675	9680	0	1	1	2	2	3	3	4	4
93	9685	9689	9694	9699	9703	9708	9713	9717	9722	9727	0	1	1	2	2	3	3	4	4
94	9731	9736	9741	9745	9750	9754	9759	9763	9768	9773	0	1	1	2	2	3	3	4	4
95	9777	9782	9786	9791	9795	9800	9805	9809	9814	9818	0	1	1	2	2	3	3	4	4
96	9823	9827	9832	9836	9841	9845	9850	9854	9859	9863	0	1	1	2	2	3	3	4	4
97	9868	9872	9877	9881	9886	9890	9894	9899	9903	9908	0	1	1	2	2	3	3	4	4
98	9912	9917	9921	9926	9930	9934	9939	9943	9948	9952	0	1	1	2	2	3	3	3	4
99	9956	9961	9965	9969	9974	9978	9983	9987	9991	9996	0	1	1	2	2	3	3	3	4
N	0	1	2	3	4	5	6	7	8	9	1	2	3	4	5	6	7	8	9

APPENDIX 3

Values and Logarithms of Exponential Functions

x	e^x Value	e^x Log_{10}	e^{-x} Value	x	e^x Value	e^x Log_{10}	e^{-x} Value
0.00	1.0000	0.00000	1.00000	0.25	1.2840	0.10857	0.77880
0.01	1.0101	0.00434	0.99005	0.26	1.2969	0.11292	0.77105
0.02	1.0202	0.00869	0.98020	0.27	1.3100	0.11726	0.76338
0.03	1.0305	0.01303	0.97045	0.28	1.3231	0.12160	0.75578
0.04	1.0408	0.01737	0.96079	0.29	1.3364	0.12595	0.74826
0.05	1.0513	0.02171	0.95123	0.30	1.3499	0.13029	0.74082
0.06	1.0618	0.02606	0.94176	0.31	1.3634	0.13463	0.73345
0.07	1.0725	0.03040	0.93239	0.32	1.3771	0.13897	0.72615
0.08	1.0833	0.03474	0.92312	0.33	1.3910	0.14332	0.71892
0.09	1.0942	0.03909	0.91393	0.34	1.4049	0.14766	0.71177
0.10	1.1052	0.04343	0.90484	0.35	1.4191	0.15200	0.70469
0.11	1.1163	0.04777	0.89583	0.36	1.4333	0.15635	0.69768
0.12	1.1275	0.05212	0.88692	0.37	1.4477	0.16069	0.69073
0.13	1.1388	0.05646	0.87809	0.38	1.4623	0.16503	0.68386
0.14	1.1503	0.06080	0.86936	0.39	1.4770	0.16937	0.67706
0.15	1.1618	0.06514	0.86071	0.40	1.4918	0.17372	0.67032
0.16	1.1735	0.06949	0.85214	0.41	1.5068	0.17806	0.66365
0.17	1.1853	0.07383	0.84366	0.42	1.5220	0.18240	0.65705
0.18	1.1972	0.07817	0.83527	0.43	1.5373	0.18675	0.65051
0.19	1.2092	0.08252	0.82696	0.44	1.5527	0.19109	0.64404
0.20	1.2214	0.08686	0.81873	0.45	1.5683	0.19543	0.63763
0.21	1.2337	0.09120	0.81058	0.46	1.5841	0.19978	0.63128
0.22	1.2461	0.09554	0.80252	0.47	1.6000	0.20412	0.62500
0.23	1.2586	0.09989	0.79453	0.48	1.6161	0.20846	0.61878
0.24	1.2712	0.10423	0.78663	0.49	1.6323	0.21280	0.61263

Note: If $0 < x < 0.01$ the value for e^{-x} can be found by the use of $(1 - x)$ or the value for e^x can be found by the use of $(1 + x)$.

x	e^x Value	Log_{10}	e^{-x} Value	x	e^x Value	Log_{10}	e^{-x} Value
0.50	1.6487	0.21715	0.60653	0.95	2.5857	0.41258	0.38674
0.51	1.6653	0.22149	0.60050	0.96	2.6117	0.41692	0.38289
0.52	1.6820	0.22583	0.59452	0.97	2.6379	0.42127	0.37908
0.53	1.6989	0.23018	0.58860	0.98	2.6645	0.42561	0.37531
0.54	1.7160	0.23452	0.58275	0.99	2.6912	0.42995	0.37158
0.55	1.7333	0.23886	0.57695	1.00	2.7183	0.43429	0.36788
0.56	1.7507	0.24320	0.57121	1.01	2.7456	0.43864	0.36422
0.57	1.7683	0.24755	0.56553	1.02	2.7732	0.44298	0.36060
0.58	1.7860	0.25189	0.55990	1.03	2.8011	0.44732	0.35701
0.59	1.8040	0.25623	0.55433	1.04	2.8292	0.45167	0.35345
0.60	1.8221	0.26058	0.54881	1.05	2.8577	0.45601	0.34994
0.61	1.8404	0.26492	0.54335	1.06	2.8864	0.46035	0.34646
0.62	1.8589	0.26926	0.53794	1.07	2.9154	0.46470	0.34301
0.63	1.8776	0.27361	0.53259	1.08	2.9447	0.46904	0.33960
0.64	1.8965	0.27795	0.52729	1.09	2.9743	0.47338	0.33622
0.65	1.9155	0.28229	0.52205	1.10	3.0042	0.47772	0.33287
0.66	1.9348	0.28664	0.51685	1.11	3.0344	0.48207	0.32956
0.67	1.9542	0.29098	0.51171	1.12	3.0649	0.48641	0.32628
0.68	1.9739	0.29532	0.50662	1.13	3.0957	0.49075	0.32303
0.69	1.9937	0.29966	0.50158	1.14	3.1268	0.49510	0.31982
0.70	2.0138	0.30401	0.49659	1.15	3.1582	0.49944	0.31664
0.71	2.0340	0.30835	0.49164	1.16	3.1899	0.50378	0.31349
0.72	2.0544	0.31269	0.48675	1.17	3.2220	0.50812	0.31037
0.73	2.0751	0.31703	0.48191	1.18	3.2544	0.51247	0.30728
0.74	2.0959	0.32138	0.47711	1.19	3.2871	0.51681	0.30422
0.75	2.1170	0.32572	0.47237	1.20	3.3201	0.52115	0.30119
0.76	2.1383	0.33006	0.46767	1.21	3.3535	0.52550	0.29820
0.77	2.1598	0.33441	0.46301	1.22	3.3872	0.52984	0.29523
0.78	2.1815	0.33875	0.45841	1.23	3.4212	0.53418	0.29229
0.79	2.2034	0.34309	0.45384	1.24	3.4556	0.53853	0.28938
0.80	2.2255	0.34744	0.44933	1.25	3.4903	0.54287	0.28650
0.81	2.2479	0.35178	0.44486	1.26	3.5254	0.54721	0.28365
0.82	2.2705	0.35612	0.44043	1.27	3.5609	0.55155	0.28083
0.83	2.2933	0.36046	0.43605	1.28	3.5966	0.55590	0.27804
0.84	2.3164	0.36481	0.43171	1.29	3.6328	0.56024	0.27527
0.85	2.3396	0.36915	0.42741	1.30	3.6693	0.56458	0.27253
0.86	2.3632	0.37349	0.42316	1.31	3.7062	0.56893	0.26982
0.87	2.3869	0.37784	0.41895	1.32	3.7434	0.57327	0.26714
0.88	2.4109	0.38218	0.41478	1.33	3.7810	0.57761	0.26448
0.89	2.4351	0.38652	0.41066	1.34	3.8190	0.58195	0.26185
0.90	2.4596	0.39087	0.40657	1.35	3.8574	0.58630	0.25924
0.91	2.4843	0.39521	0.40252	1.36	3.8962	0.59064	0.25666
0.92	2.5093	0.39955	0.39852	1.37	3.9354	0.59498	0.25411
0.93	2.5345	0.40389	0.39455	1.38	3.9749	0.59933	0.25158
0.94	2.5600	0.40824	0.39063	1.39	4.0149	0.60367	0.24908

x	e^x Value	Log_{10}	e^{-x} Value	x	e^x Value	Log_{10}	e^{-x} Value
1.40	4.0552	0.60801	0.24660	1.85	6.3598	0.80344	0.15724
1.41	4.0960	0.61236	0.24414	1.86	6.4237	0.80779	0.15567
1.42	4.1371	0.61670	0.24171	1.87	6.4883	0.81213	0.15412
1.43	4.1787	0.62104	0.23931	1.88	6.5535	0.81647	0.15259
1.44	4.2207	0.62538	0.23693	1.89	6.6194	0.82082	0.15107
1.45	4.2631	0.62973	0.23457	1.90	6.6859	0.82516	0.14957
1.46	4.3060	0.63407	0.23224	1.91	6.7531	0.82950	0.14808
1.47	4.3492	0.63841	0.22993	1.92	6.8210	0.83385	0.14661
1.48	4.3929	0.64276	0.22764	1.93	6.8895	0.83819	0.14515
1.49	4.4371	0.64710	0.22537	1.94	6.9588	0.84253	0.14370
1.50	4.4817	0.65144	0.22313	1.95	7.0287	0.84687	0.14227
1.51	4.5267	0.65578	0.22091	1.96	7.0993	0.85122	0.14086
1.52	4.5722	0.66013	0.21871	1.97	7.1707	0.85556	0.13946
1.53	4.6182	0.66447	0.21654	1.98	7.2427	0.85990	0.13807
1.54	4.6646	0.66881	0.21438	1.99	7.3155	0.86425	0.13670
1.55	4.7115	0.67316	0.21225	2.00	7.3891	0.86859	0.13534
1.56	4.7588	0.67750	0.21014	2.01	7.4633	0.87293	0.13399
1.57	4.8066	0.68184	0.20805	2.02	7.5383	0.87727	0.13266
1.58	4.8550	0.68619	0.20598	2.03	7.6141	0.88162	0.13134
1.59	4.9037	0.69053	0.20393	2.04	7.6906	0.88596	0.13003
1.60	4.9530	0.69487	0.20190	2.05	7.7679	0.89030	0.12873
1.61	5.0028	0.69921	0.19989	2.06	7.8460	0.89465	0.12745
1.62	5.0531	0.70356	0.19790	2.07	7.9248	0.89899	0.12619
1.63	5.1039	0.70790	0.19593	2.08	8.0045	0.90333	0.12493
1.64	5.1552	0.71224	0.19398	2.09	8.0849	0.90768	0.12369
1.65	5.2070	0.71659	0.19205	2.10	8.1662	0.91202	0.12246
1.66	5.2593	0.72093	0.19014	2.11	8.2482	0.91636	0.12124
1.67	5.3122	0.72527	0.18825	2.12	8.3311	0.92070	0.12003
1.68	5.3656	0.72961	0.18637	2.13	8.4149	0.92505	0.11884
1.69	5.4195	0.73396	0.18452	2.14	8.4994	0.92939	0.11765
1.70	5.4739	0.73830	0.18268	2.15	8.5849	0.93373	0.11648
1.71	5.5290	0.74264	0.18087	2.16	8.6711	0.93808	0.11533
1.72	5.5845	0.74699	0.17907	2.17	8.7583	0.94242	0.11418
1.73	5.6407	0.75133	0.17728	2.18	8.8463	0.94676	0.11304
1.74	5.6973	0.75567	0.17552	2.19	8.9352	0.95110	0.11192
1.75	5.7546	0.76002	0.17377	2.20	9.0250	0.95545	0.11080
1.76	5.8124	0.76436	0.17204	2.21	9.1157	0.95979	0.10970
1.77	5.8709	0.76870	0.17033	2.22	9.2073	0.96413	0.10861
1.78	5.9299	0.77304	0.16864	2.23	9.2999	0.96848	0.10753
1.79	5.9895	0.77739	0.16696	2.24	9.3933	0.97282	0.10646
1.80	6.0496	0.78173	0.16530	2.25	9.4877	0.97716	0.10540
1.81	6.1104	0.78607	0.16365	2.26	9.5831	0.98151	0.10435
1.82	6.1719	0.79042	0.16203	2.27	9.6794	0.98585	0.10331
1.83	6.2339	0.79476	0.16041	2.28	9.7767	0.99019	0.10228
1.84	6.2965	0.79910	0.15882	2.29	9.8749	0.99453	0.10127

x	e^x Value	e^x Log$_{10}$	e^{-x} Value	x	e^x Value	e^x Log$_{10}$	e^{-x} Value
2.30	9.9742	0.99888	0.10026	2.75	15.643	1.19431	0.06393
2.31	10.074	1.00322	0.09926	2.76	15.800	1.19865	0.06329
2.32	10.176	1.00756	0.09827	2.77	15.959	1.20300	0.06266
2.33	10.278	1.01191	0.09730	2.78	16.119	1.20734	0.06204
2.34	10.381	1.01625	0.09633	2.79	16.281	1.21168	0.06142
2.35	10.486	1.02059	0.09537	2.80	16.445	1.21602	0.06081
2.36	10.591	1.02493	0.09442	2.81	16.610	1.22037	0.06020
2.37	10.697	1.02928	0.09348	2.82	16.777	1.22471	0.05961
2.38	10.805	1.03362	0.09255	2.83	16.945	1.22905	0.05901
2.39	10.913	1.03796	0.09163	2.84	17.116	1.23340	0.05843
2.40	11.023	1.04231	0.09072	2.85	17.288	1.23774	0.05784
2.41	11.134	1.04665	0.08982	2.86	17.462	1.24208	0.05727
2.42	11.246	1.05099	0.08892	2.87	17.637	1.24643	0.05670
2.43	11.359	1.05534	0.08804	2.88	17.814	1.25077	0.05613
2.44	11.473	1.05968	0.08716	2.89	17.993	1.25511	0.05558
2.45	11.588	1.06402	0.08629	2.90	18.174	1.25945	0.05502
2.46	11.705	1.06836	0.08543	2.91	18.357	1.26380	0.05448
2.47	11.822	1.07271	0.08458	2.92	18.541	1.26814	0.05393
2.48	11.941	1.07705	0.08374	2.93	18.728	1.27248	0.05340
2.49	12.061	1.08139	0.08291	2.94	18.916	1.27683	0.05287
2.50	12.182	1.08574	0.08208	2.95	19.106	1.28117	0.05234
2.51	12.305	1.09008	0.08127	2.96	19.298	1.28551	0.05182
2.52	12.429	1.09442	0.08046	2.97	19.492	1.28985	0.05130
2.53	12.554	1.09877	0.07966	2.98	19.688	1.29420	0.05079
2.54	12.680	1.10311	0.07887	2.99	19.886	1.29854	0.05029
2.55	12.807	1.10745	0.07808	3.00	20.086	1.30288	0.04979
2.56	12.936	1.11179	0.07730	3.05	21.115	1.32460	0.04736
2.57	13.066	1.11614	0.07654	3.10	22.198	1.34631	0.04505
2.58	13.197	1.12048	0.07577	3.15	23.336	1.36803	0.04285
2.59	13.330	1.12482	0.07502	3.20	24.533	1.38974	0.04076
2.60	13.464	1.12917	0.07427	3.25	25.790	1.41146	0.03877
2.61	13.599	1.13351	0.07353	3.30	27.113	1.43317	0.03688
2.62	13.736	1.13785	0.07280	3.35	28.503	1.45489	0.03508
2.63	13.874	1.14219	0.07208	3.40	29.964	1.47660	0.03337
2.64	14.013	1.14654	0.07136	3.45	31.500	1.49832	0.03175
2.65	14.154	1.15088	0.07065	3.50	33.115	1.52003	0.03020
2.66	14.296	1.15522	0.06995	3.55	34.813	1.54175	0.02872
2.67	14.440	1.15957	0.06925	3.60	36.598	1.56346	0.02732
2.68	14.585	1.16391	0.06856	3.65	38.475	1.58517	0.02599
2.69	14.732	1.16825	0.06788	3.70	40.447	1.60689	0.02472
2.70	14.880	1.17260	0.06721	3.75	42.521	1.62860	0.02352
2.71	15.029	1.17694	0.06654	3.80	44.701	1.65032	0.02237
2.72	15.180	1.18128	0.06587	3.85	46.993	1.67203	0.02128
2.73	15.333	1.18562	0.06522	3.90	49.402	1.69375	0.02024
2.74	15.487	1.18997	0.06457	3.95	51.935	1.71546	0.01925

x	e^x Value	e^x Log$_{10}$	e^{-x} Value	x	e^x Value	e^x Log$_{10}$	e^{-x} Value
4.00	54.598	1.73718	0.01832	5.50	244.69	2.38862	0.00409
4.10	60.340	1.78061	0.01657	5.60	270.43	2.43205	0.00370
4.20	66.686	1.82404	0.01500	5.70	298.87	2.47548	0.00335
4.30	73.700	1.86747	0.01357	5.80	330.30	2.51891	0.00303
4.40	81.451	1.91090	0.01227	5.90	365.04	2.56234	0.00274
4.50	90.017	1.95433	0.01111	6.00	403.43	2.60577	0.00248
4.60	99.484	1.99775	0.01005	6.25	518.01	2.71434	0.00193
4.70	109.95	2.04118	0.00910	6.50	665.14	2.82291	0.00150
4.80	121.51	2.08461	0.00823	6.75	854.06	2.93149	0.00117
4.90	134.29	2.12804	0.00745	7.00	1096.6	3.04006	0.00091
5.00	148.41	2.17147	0.00674	7.50	1808.0	3.25721	0.00055
5.10	164.02	2.21490	0.00610	8.00	2981.0	3.47436	0.00034
5.20	181.27	2.25833	0.00552	8.50	4914.8	3.69150	0.00020
5.30	200.34	2.30176	0.00499	9.00	8103.1	3.90865	0.00012
5.40	221.41	2.34519	0.00452	9.50	13360.	4.12580	0.00007
				10.00	22026.	4.34294	0.00005

APPENDIX 4

Three-place values of Trigonometric Functions and Degrees in Radian Measure

Rad.	Deg.	Sin	Tan	Sec	Csc	Cot	Cos	Deg.	Rad.
.000	0°	0.000	0.000	1.000	——	——	1.000	90°	1.571
.017	1°	0.017	0.017	1.000	57.30	57.29	1.000	89°	1.553
.035	2°	0.035	0.035	1.001	28.65	28.64	0.999	88°	1.536
.052	3°	0.052	0.052	1.001	19.11	19.08	.999	87°	1.518
.070	4°	0.070	0.070	1.002	14.34	14.30	.998	86°	1.501
.087	5°	0.087	0.087	1.004	11.47	11.43	.996	85°	1.484
.105	6°	0.105	0.105	1.006	9.567	9.514	.995	84°	1.466
.122	7°	0.122	0.123	1.008	8.206	8.144	.993	83°	1.449
.140	8°	0.139	0.141	1.010	7.185	7.115	.990	82°	1.431
.157	9°	0.156	0.158	1.012	6.392	6.314	.988	81°	1.414
.175	10°	0.174	0.176	1.015	5.759	5.671	.985	80°	1.396
.192	11°	0.191	0.194	1.019	5.241	5.145	.982	79°	1.379
.209	12°	0.208	0.213	1.022	4.810	4.705	.978	78°	1.361
.227	13°	0.225	0.231	1.026	4.445	4.331	.974	77°	1.344
.244	14°	0.242	0.249	1.031	4.134	4.011	.970	76°	1.326
.262	15°	0.259	0.268	1.035	3.864	3.732	.966	75°	1.309
.279	16°	0.276	0.287	1.040	3.628	3.487	.961	74°	1.292
.297	17°	0.292	0.306	1.046	3.420	3.271	.956	73°	1.274
.314	18°	0.309	0.325	1.051	3.236	3.078	.951	72°	1.257
.332	19°	0.326	0.344	1.058	3.072	2.904	.946	71°	1.239
.349	20°	0.342	0.364	1.064	2.924	2.747	.940	70°	1.222
.367	21°	0.358	0.384	1.071	2.790	2.605	.934	69°	1.204
.384	22°	0.375	0.404	1.079	2.669	2.475	.927	68°	1.187
.401	23°	0.391	0.424	1.086	2.559	2.356	.921	67°	1.169
.419	24°	0.407	0.445	1.095	2.459	2.246	.914	66°	1.152
.436	25°	0.423	0.466	1.103	2.366	2.145	.906	65°	1.134
Rad.	Deg.	Cos	Cot	Csc	Sec	Tan	Sin	Deg.	Rad.

Rad.	Deg.	Sin	Tan	Sec	Csc	Cot	Cos	Deg.	Rad.
.454	26°	0.438	0.488	1.113	2.281	2.050	.899	64°	1.117
.471	27°	0.454	0.510	1.122	2.203	1.963	.891	63°	1.100
.489	28°	0.469	0.532	1.133	2.130	1.881	.883	62°	1.082
.506	29°	0.485	0.554	1.143	2.063	1.804	.875	61°	1.065
.524	30°	0.500	0.577	1.155	2.000	1.732	.866	60°	1.047
.541	31°	0.515	0.601	1.167	1.942	1.664	.857	59°	1.030
.559	32°	0.530	0.625	1.179	1.887	1.600	.848	58°	1.012
.576	33°	0.545	0.649	1.192	1.836	1.540	.839	57°	0.995
.593	34°	0.559	0.675	1.206	1.788	1.483	.829	56°	0.977
.611	35°	0.574	0.700	1.221	1.743	1.428	.819	55°	0.960
.628	36°	0.588	0.727	1.236	1.701	1.376	.809	54°	0.942
.646	37°	0.602	0.754	1.252	1.662	1.327	.799	53°	0.925
.663	38°	0.616	0.781	1.269	1.624	1.280	.788	52°	0.908
.681	39°	0.629	0.810	1.287	1.589	1.235	.777	51°	0.890
.698	40°	0.643	0.839	1.305	1.556	1.192	.766	50°	0.873
.716	41°	0.656	0.869	1.325	1.524	1.150	.755	49°	0.855
.733	42°	0.669	0.900	1.346	1.494	1.111	.743	48°	0.838
.750	43°	0.682	0.933	1.367	1.466	1.072	.731	47°	0.820
.768	44°	0.695	0.966	1.390	1.440	1.036	.719	46°	0.803
.785	45°	0.707	1.000	1.414	1.414	1.000	.707	45°	0.785
Rad.	Deg.	Cos	Cot	Csc	Sec	Tan	Sin	Deg.	Rad.

TYPE SN7490N
DECADE COUNTER

absolute maximum ratings over operating free-air temperature range (unless otherwise noted)

Supply Voltage V_{CC} (See Note 3) .... 7 V
Input Voltage V_{in} (See Notes 3 and 4) .... 5.5 V
Operating Free-Air Temperature Range .... 0°C to 70°C
Storage Temperature Range .... −55°C to 125°C

NOTES: 3. These voltage values are with respect to network ground terminal.
4. Input signals must be zero or positive with respect to network ground terminal.

recommended operating conditions

Supply Voltage V_{CC} .... 4.75 V to 5.25 V
Fan-Out From Each Output (See Note 5) .... 1 to 10
Width of Input Count Pulse, $t_{p(in)}$ .... ≥ 50 ns
Width of Reset Pulse, $t_{p(reset)}$ .... ≥ 50 ns

NOTE 5: Fan-out from output A to input BD and to 10 additional Series 74 loads is permitted.

electrical characteristics, $T_A = 0°C$ to 70°C

PARAMETER		TEST FIG.	TEST CONDITIONS	MIN	TYP	MAX	UNIT
$V_{in(1)}$	Input voltage required to ensure logical 1 at inputs A, $R_{0(1)}$, $R_{0(2)}$, $R_{9(1)}$, and $R_{9(2)}$	1	$V_{CC} = 4.75$ V	2			V
$V_{in(1)}$	Input voltage required to ensure logical 1 at input BD	1	$V_{CC} = 4.75$ V	2.2			V
$V_{in(0)}$	Input voltage required to ensure logical 0 at inputs A, $R_{0(1)}$, $R_{0(2)}$, $R_{9(1)}$, and $R_{9(2)}$	2	$V_{CC} = 4.75$ V			0.8	V
$V_{in(0)}$	Input voltage required to ensure logical 0 at input BD	2	$V_{CC} = 4.75$ V			0.6	V
$V_{out(1)}$	Logical 1 output voltage	2	$V_{CC} = 4.75$ V, $I_{load} = -400 \mu A$	2.4			V

PARAMETER		TEST FIG.	TEST CONDITIONS	MIN	TYP	MAX	UNIT
$V_{out(0)}$	Logical 0 output voltage	1	$V_{CC}=4.75$ V, $I_{sink}=16$ mA			0.4	V
$I_{in(1)}$	Logical 1 level input current at $R_{0(1)}$, $R_{W(2)}$, $R_{9(1)}$, or $R_{9(2)}$	3	$V_{CC}=5.25$ V, $V_{in}=2.4$ V			40	μA
			$V_{CC}=5.25$ V, $V_{in}=5.5$ V			1	mA
$I_{in(1)}$	Logical 1 level input current at input A	3	$V_{CC}=5.25$ V, $V_{in}=2.4$ V			80	μA
			$V_{CC}=5.25$ V, $V_{in}=5.5$ V			1	mA
$I_{in(1)}$	Logical 1 level input current at input BD	3	$V_{CC}=5.25$ V, $V_{in}=2.4$ V			160	μA
			$V_{CC}=5.25$ V, $V_{in}=5.5$ V			1	mA
$I_{in(0)}$	Logical 0 level input current at $R_{0(1)}$, $R_{W(2)}$, $R_{9(1)}$, or $R_{9(2)}$	4	$V_{CC}=5.25$ V, $V_{in}=0.4$ V			-1.6	mA
$I_{in(0)}$	Logical 0 level input current at input A	4	$V_{CC}=5.25$ V, $V_{in}=0.4$ V			-3.2	mA
$I_{in(0)}$	Logical 0 level input current at input BD	4	$V_{CC}=5.25$ V, $V_{in}=0.4$ V			-6.4	mA
I_{OS}	Short-circuit output current†	5	$V_{CC}=5.25$ V, $V_{out}=0$V	-18		-57	mA
I_{CC}	Supply Current	3	$V_{CC}=5$ V, $T_A=25°C$		32		mA

† Not more than one output should be shorted at a time.

switching characteristics, $V_{CC}=5$ V, $T_A=25°C$, N = 10

PARAMETER	TEST FIG.	TEST CONDITIONS	MIN	TYP	MAX	UNIT
f_{max}	Maximum frequency of input count pulses			10	18		MHz
t_{pd1}	Propagation delay time to logical 1 level from input count pulse to output C	6			60	100	ns
t_{pd0}	Propagation delay time to logical 0 level from input count pulse to output C	6			60	100	ns

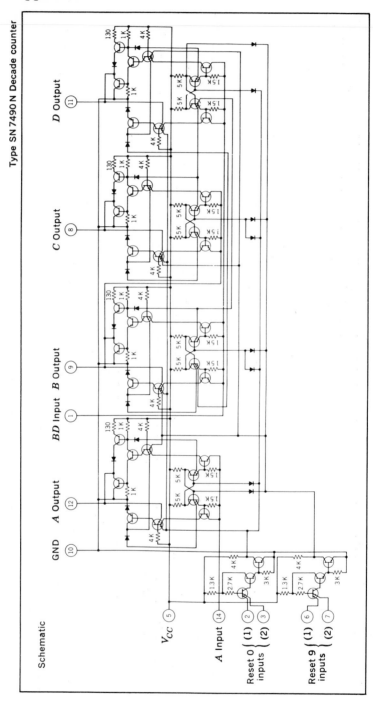

Type SN 7490 N Decade counter

Schematic

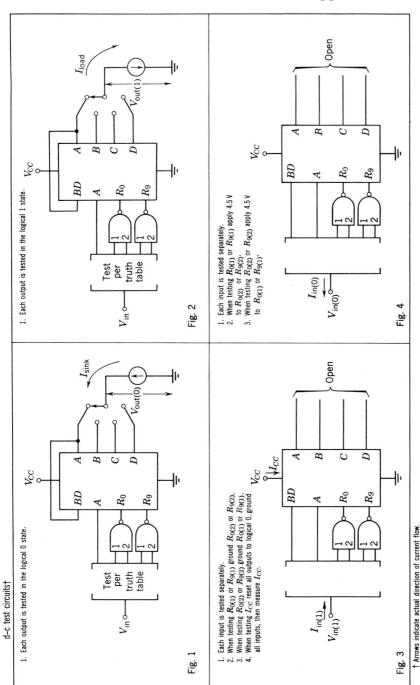

d-c test circuits†

Fig. 1

1. Each output is tested in the logical 0 state.

1. Each input is tested separately.
2. When testing $R_{0(1)}$ or $R_{9(1)}$ ground $R_{0(2)}$ or $R_{9(2)}$.
3. When testing $R_{0(2)}$ or $R_{9(2)}$ ground $R_{0(1)}$ or $R_{9(1)}$.
4. When testing I_{CC} reset all outputs to logical 0, ground all inputs, then measure I_{CC}.

Fig. 3

1. Each output is tested in the logical 1 state.

Fig. 2

1. Each input is tested separately.
2. When testing $R_{0(1)}$ or $R_{9(1)}$ apply 4.5 V to $R_{0(2)}$ or $R_{9(2)}$.
3. When testing $R_{0(2)}$ or $R_{9(2)}$ apply 4.5 V to $R_{0(1)}$ or $R_{9(1)}$.

Fig. 4

† Arrows indicate actual direction of current flow.

332 *Appendix 5*

d–c test circuits †
(continued)

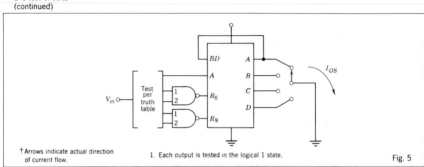

† Arrows indicate actual direction of current flow.

1. Each output is tested in the logical 1 state.

Fig. 5

switching time voltage waveforms

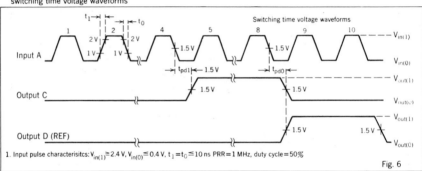

1. Input pulse characterisitcs: $V_{in(1)} \geq 2.4$ V, $V_{in(0)} \leq 0.4$ V, $t_1 = t_0 \leq 10$ ns PRR=1 MHz, duty cycle=50%

Fig. 6

mechanical data

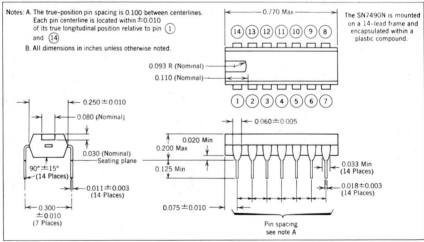

Notes: A. The true–position pin spacing is 0.100 between centerlines. Each pin centerline is located within ±0.010 of its true longitudinal position relative to pin ① and ⑭.

B. All dimensions in inches unless otherwise noted.

The SN7490N is mounted on a 14-lead frame and encapsulated within a plastic compound.

APPENDIX 6

TYPE SN7441N
BCD-TO-DECIMAL DECODER/DRIVER

recommended operating conditions

Supply Voltage V_{CC} . . . . 4.75 V to 5.25 V
Maximum Voltage on Any output . . . . 55 V

absolute maximum ratings over operating free-air temperature range unless otherwise noted

Supply Voltage V_{CC} (See Note 1) . . . . 7 V
Input Voltage, V_{in} (See Notes 1 and 2) . . . . 5.5 V
Current Into Any Output (Off-State) . . . . 0.5 mA
Operating Free-air Temperature Range . . . . 0°C to 70°C
Storage Temperature Range . . . . −55°C to 125°C

NOTES: 1. Voltage values are with respect to network ground terminal.
2. Input signals must be zero or positive with respect to network ground terminal.

electrical characteristics, $T_A = 0°C$ to 70°C

	PARAMETER	TEST FIG.		MIN	TYP	MAX	UNIT
$V_{in(1)}$	Logical 1 input voltage	1	$V_{CC} = 4.75$ V	2			V
$V_{in(0)}$	Logical 0 input voltage	1	$V_{CC} = 4.75$ V			0.8	V
V_{on}	On-state output voltage	1	$V_{CC} = 4.75$ V, $I_{sink} = 7$ mA			2.5	V
I_{OL}	Output leakage current	2	$V_{CC} = 5.25$ V, $V_{out} = 55$ V			200	μA
$I_{in(1)}$	Logical 1 level input current at B, C, or D	3	$V_{CC} = 5.25$ V, $V_{in} = 2.4$ V			40	μA
			$V_{CC} = 5.25$ V, $V_{in} = 5.5$ V			1	mA
$I_{in(1)}$	Logical 1 level input current at A	3	$V_{CC} = 5.25$ V, $V_{in} = 2.4$ V			80	μA
			$V_{CC} = 5.25$ V, $V_{in} = 5.5$ V			1	mA
$I_{in(0)}$	Logical 0 level input current at B, C, or D	4	$V_{CC} = 5.25$ V, $V_{in} = 0.4$ V			−1.6	mA
$I_{in(0)}$	Logical 0 level input current at A	4	$V_{CC} = 5.25$ V, $V_{in} = 0.4$ V			−3.2	mA
I_{CC}	Supply current	3	$V_{CC} = 5$ V, $T_A = 25°C$		21		mA

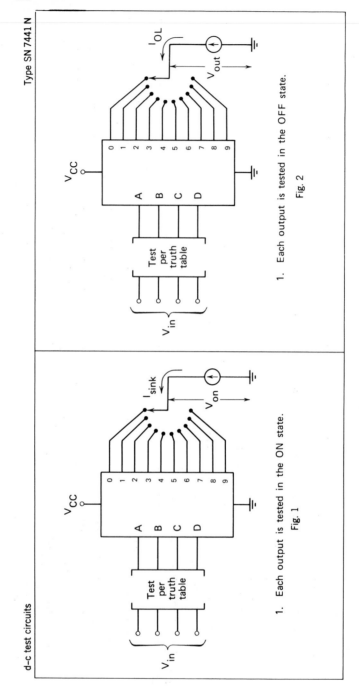

d-c test circuits

Type SN 7441 N

V_{CC}

0
1
2
3
4
5
6
7
8
9

A
B
C
D

Test
per
truth
table

V_{in}

I_{OL}

V_{out}

1. Each output is tested in the OFF state.

Fig. 2

V_{CC}

0
1
2
3
4
5
6
7
8
9

A
B
C
D

Test
per
truth
table

V_{in}

I_{sink}

V_{on}

1. Each output is tested in the ON state.

Fig. 1

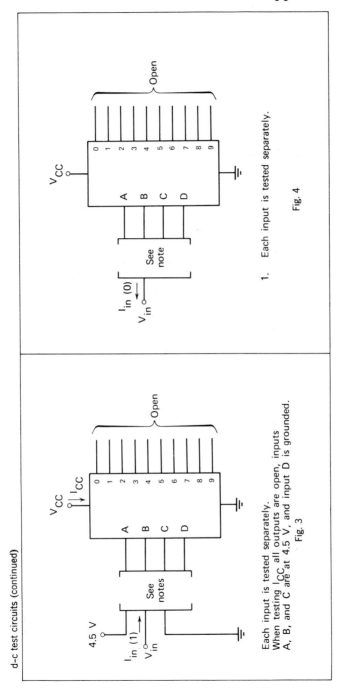

d-c test circuits (continued)

V_{CC} I_{CC}

A
B
C
D

0
1
2
3
4
5
6
7
8
9

Open

See notes

4.5 V

I_{in} (1)
V_{in}

Each input is tested separately.
When testing I_{CC} all outputs are open, inputs
A, B, and C are at 4.5 V, and input D is grounded.

Fig. 3

V_{CC}

A
B
C
D

0
1
2
3
4
5
6
7
8
9

Open

See note

I_{in} (0)
V_{in}

1. Each input is tested separately.

Fig. 4

TYPE SN7441N
typical application

It is recommended that switching be accomplished with a specified
prebias voltage (V_{PB}) applied to the cathode of the indicator tube (see Fig. 5).

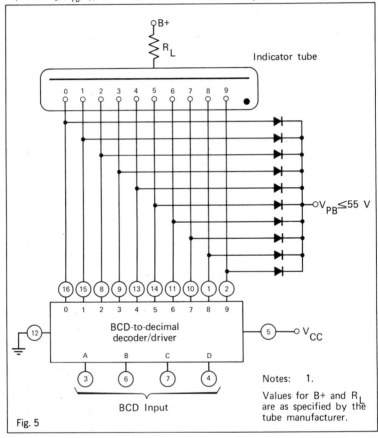

Indicator tube

$V_{PB} \leq 55$ V

BCD-to-decimal
decoder/driver

V_{CC}

BCD Input

Notes: 1.

Values for B+ and R_L
are as specified by the
tube manufacturer.

Fig. 5

Mechanical data

The SN7441N circuit is mounted on a 16-lead frame and
encapsulated within a plastic compound.

Notes: A. The true-position pin spacing is 0.100 between centerlines. Each pin centerline is located within ± 0.010
of its true longitudinal position relative to pin ①
and ⑯

B. All dimensions in inches unless otherwise noted.

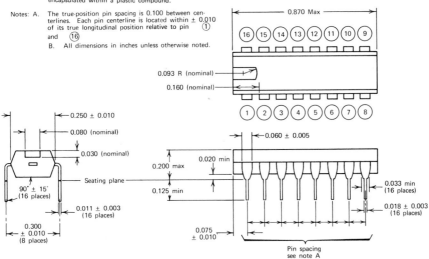

INDEX

Adders, 146–150
Adder-subtractor, *see* Binary adders
Address, 218
Addressing and sense wires, 253
Amplifiers, 80–88
  voltage feedback, 80
Analog-to-digital converter (ADC), 216–
  218
Anticoincidence circuit, 111, 199, 204
Antiwalk characteristics, 203
Aspect ratio, 169
Astable multivibrator, 120
Attenuator, 170
Avalanche diode, *see* Zener diode

Background count, 261
Base line, 43
BCD-to-decimal decoder, 307
Bilateral transistor, 229
Binary adders, 146–150
Binary subtractor, *see* Half adder-subtractor
Bistable multivibrator, *see* Flip-flop
Blocking oscillator, 121
Boltzmann constant, 45
Boolean algebra, 100

Bootstrap mechanism, 239
Bootstrapped circuit, 160, 181
Breakdown voltage, transistor, 69

Campbell's theorem, 163
Capacitance, definition, 2
  parasitic, 158
Carrier-activation noise, 184
Cascode pair, 184
Cathode-ray-tube circuit, 298
Clamping circuit, 52–60
  theorem, 56, 59
Class A operation, 188
Clipping circuit, 52
Clock pulse, 102
Coincidence circuits, 106–110
Common-mode component, 86
Complementary emitter-follower, 88
Computer, 100
Counter, 135–141
  decade (decimal), with feedback, 139
    without feedback, 141
  a four-binary chain, 135
Current amplifier, 156
Current feedback, 83

Current-sensitive amplifier, 181
Cut-in voltage, diode, 54
    transistor, 92
Cutoff current, transistor, 68

Dc amplifier, 80–88
Dead time, 222
Decimal-to-binary conversion, 101
Delay lines, 38–43
    applications, 38
    differentiation, 41
    pulse shaping, 40
Detector characteristics, 155
Difference (differential) amplifier, 85
Digital computer, see Computer
Digital system, 100
Digital-to-analog converter (DAC), 218
Diodes, 43–62
    breakdown, 49
    circuits, 51–62
    p-n junction, 45
    semiconductor, 44
    thermionic, 43
    tunnel, 49
    Zener, 47
Discriminator, lower-level, 199
    phase, 267
    upper-level, 199
Double diode clipper, 52
Duty cycle, 166
    effect, 160
Dwell timing, 235
Dynamic capacitor, 266

Eccles-Jordan type multivibrator, 126

Feedback loops, 181
Ferrite core memories, 227, 252
Field-effect transistor (FET), 71
Flip-flop, 125–135
    Eccles-Jordan type, 126
    external characteristics, 116
    principal characteristics, 125
    self-biased, 126
    symmetrical triggering, 130–133
    unsymmetrical triggering, 133–135
Frequency cutoff, transistor, 68
Full-wave rectifiers, bridge circuit, 46
    center-tapped transformer, 46
Full width at half maximum (FWHM), 170

Gain-bandwidth product, transistor, 68
Geiger-Müller tube, 259
Ground-loop noise, 171

Half adder-subtractor, 148
Half-wave rectification, 43
Half-wave rectifiers as power supplies, 60–62
High-pass RC network, 5–16; see also RC
    network, high-pass
Hybrid parameters, transistor, 75–77
Hysteresis of a regenerative trigger, 142, 203

Inductance, definition, 2
    leakage, 29
    magnetizing, 29
    measurement, 31
Inhibit-pulse generator, 255
Inhibit wire, 230
Ionization chamber, 266

Linear gate circuit, 248
Linearity, differential, 222
    integral, 222
Live time, 221
Logic circuits, 100–116
    AND gate (coincidence circuit), 107–110
    EXCLUSIVE OR gate, 113
    flip-flop, 116
    inhibitor (anticoincidence circuit), 111
    logic system, 100–103
    NAND gate, 114
    NOR gate, 116
    NOT gate (inverter), 111
    OR gate, 103–106
Low-pass RC network, see RC network, low-
    pass

Memory cycle, 255
Memory register, 254
Miller integrator circuit, 290
Miller's theorem, 80–82
Milliroentgen/hr (mR/hr), 259
Monostable multivibrator, 144–146
Multichannel analyzer, multiparameter, 223–
    225
    one parameter, 216
Multivibrator, astable, 120
    bistable, see Flip-flop
    monostable, 144–146

Negative resistance, 50
Network theorems, 4–5
  Kirchhoff's laws, 4
  Norton's theorem, 5
  superposition theorem, 4
  Thevenin's theorem, 5
Nuvistor cathode-follower, 282

One-shot multivibrator, 144
Open-circuit impedance parameters, transistor, 73–74
Operational amplifier, 80–85
  differentiator, 84–85
  integrator, 83–84
  principle, 80–83

Parallel binary adder, 143
Peaking circuit, 24
Pentupler circuit, 298
Phase discriminator, 267
Pileup, 157, 162
Pole-zero cancellation, 183
Preamplifiers, 173–180
  BF_3 counting system, 175
  cathode- (emitter-) follower, 173
  charge-sensitive, 183
  Darlington emitter-follower, 175
  integrator, 180
  photomultiplier, 175
  White cathode-follower, 175
Proportional type detector, 272–273
Pulse-height-to-time converter, see Analog-to-digital converter
Pulse transformers, 28–38
  capacitance, 29
  complete pulse response of, 35
  flat-top response of, 33
  rise time response of, 31
  winding-and-core considerations, 37

RC network, 5–23
  high-pass, 5–16
    differentiator, 13
    double differentiation, 15
    exponential input, 11
    pulse input, 7
    sine-wave input, 5
    square-wave input, 7
    step-function input, 5
    sweep (ramp) input, 13

  low-pass, 16–23
    exponential input, 20
    integrator, 22
    pulse input, 18
    sine-wave input, 16
    square-wave input, 19
    step-function input, 16
    sweep (ramp) input, 22
Read cycle, 229
Reserve gain, 184
Reset to zero condition, 310
Residual magnetism, 252
Resistance, definition, 1
  negative, 50
Resolving time, 161
Resonance, series, 3
  parallel, 4
Reverse saturation current, 45
Ringing circuit, 27
Rise time, definition, 17
RLC circuit, 23

Sampling gate, 110
Saturation level, 95
Scaler, binary, 139
  decade, see Decade counter
  integrated circuit, 308–313
Schmitt trigger circuit, 141
Semiconductor diode, 44
Sense amplifier, 229, 256
Serial binary adder, 150
Short-circuit admittance parameters, transistor, 74
Speed-up capacitor, 126
Stability control, 294
Stretcher, 205
Switching circuit principle, 88–93

Tail pulse, 157
Temperature-stable components, 173
Thermionic diode, 44
Three regions of operation, transistor, 70
Threshold voltage level, 201
Time-base generator, 290
Time-base trigger, 286
Time-to-digital converter (TDC), 218
Transconductance, transistor, 68
Transient waveform, 54
Transistor circuits, common base, 64–66
  common collector, 67

common emitter, 64–67
Transistor oscillators, 117–125
  blocking, 121
  multivibrator, 118–120
  nonsinusoidal, 118
  sinusoidal, 117–118
Transistors, 64–71
Transistor switch, 93
Transmission gate, 110
Transmission line, *see* Delay lines
Triode characteristics, 62–64
Tunnel diode, 49

Unblanking, 295

Virtual ground, 80
Voltage feedback, 80
Voltage multipliers, doubler, 46
  quadrupler, 47
  tripler, 47
Voltage window, 199

Word, 101
Write cycle, 229, 254

Zener diode, 47–49
Zero drift, 222